D0734595

55
3

HISTORY
OF
LIFE

McGRAW-HILL EARTH SCIENCE PAPERBACK SERIES

Richard Ojakangas, Consulting Editor

Bird and Goodman: PLATE TECTONICS

Cowen: HISTORY OF LIFE

Kesler: OUR FINITE MINERAL RESOURCES

Matsch: NORTH AMERICA AND THE GREAT ICE AGE

Oakeshott: VOLCANOES AND EARTHQUAKES: GEOLOGIC VIOLENCE

Ojakangas and Darby: THE EARTH AND ITS HISTORY

HISTORY OF LIFE

RICHARD COWEN
University of California, Davis

McGRAW-HILL BOOK COMPANY

New York St. Louis San Francisco Auckland Düsseldorf Johannesburg
Kuala Lumpur London Mexico Montreal New Delhi Panama
Paris São Paulo Singapore Sydney Tokyo Toronto

253810

This book was set in Helvetica by Black Dot, Inc.
The editors were Robert H. Summersgill and James R. Belser;
the designer was J. E. O'Connor;
the production supervisor was Dennis J. Conroy.
The drawings were done by J & R Services, Inc.
Kingsport Press, Inc., was printer and binder.

HISTORY OF LIFE

Copyright © 1976 by McGraw-Hill, Inc. All rights reserved.
Printed in the United States of America. No part of this publication
may be reproduced, stored in a retrieval system, or transmitted, in any
form or by any means, electronic, mechanical, photocopying, recording, or
otherwise, without the prior written permission of the publisher.

34567890KPKP7987

Library of Congress Cataloging in Publication Data

Cowen, Richard, date
    History of life

    (McGraw-Hill earth science paperback series)
    1. Paleontology. I. Title.
QE711.2.C68   560   74-26894
ISBN 0-07-013260-7 pbk.

CONTENTS

PREFACE

A TV news reporter tries to relate the events of the day which seem important to him. But so many events happen in the lives of billions of people that a lifetime of TV programs couldn't truly show the history of Earth during a 24-hour period. So certain facts are selected from a host of happenings because they seem significant at the time, particularly if they affect the lives of many people.

A reporter tries to set the events of the day into a pattern if he can see one. For example, a news item about fuel shortage might be placed next to a report about a period of frosty weather in New England, so that the viewer can see the importance of each event in relation to the other.

Sometimes the reporter may introduce a political observer who will discuss events which may have happened although there are no hard facts to go on—thus he might speculate about the progress of secret talks before any official communiqué is released. Finally, a TV news reporter tries to make his program interesting—it is useless to try to inform people about the events of the day if they are too bored to pay attention.

Writing history is much the same art as putting together a news program. One selects important facts and tries to fit them together into a pattern that makes sense. One is allowed to fill in gaps in hard knowledge with speculative opinions as long as the speculation is intelligent and based on some fund of knowledge. Hopefully, a historian tries to reconstruct events that would be of interest to someone besides himself—his aim is to inform other people, after all.

This history of life has been written with such an approach in mind. For facts we have the rock and fossil record. For intelligent speculation we can call on our knowledge of the laws of physics and chemistry which are universal in time and space, and we can look at the organisms of the past with the help of knowledge about the biology of living organisms. The story as a whole has to make sense of the facts, and it has to make sense in terms of the physical, chemical, and biological processes we are familiar with.

The history of life is fascinating because it is the unfolding story of all the living things on this Earth. The dramatic and turbulent events of their past seem to make sense in relation to the history of the Earth they inhabit, and if we can understand what made living things the way they are, perhaps we can arrange that their future will be something more than the black cloud of extinction.

Bringing life to the past is one of the greatest of highs—it is almost entirely an imaginative process. One can't simply be an observer, as one can when looking at living animals that have their own existence. To bring a fossil to life demands the investment of part of oneself. Every step a dinosaur takes is generated by the paleobiologist who reconstructs it; one has to *be* a dinosaur before one can understand their powers and their problems.

I hope that reading this book will encourage you to conjure up your individual imaginative picture of the Earth in the past. I have tried to indicate the kinds of pictures that are allowed by the facts of the rock and fossil record as we know them at present. The pictures may change as new facts are discovered, but the richness of the fossil record will always provide interest and excitement for those who like to empathize with living things, however dusty their bones may seem to be.

ACKNOWLEDGMENTS
This book evolved by natural selection in the environment of Geology 3 at the University of California, Davis. The students who were interested, excited, skeptical, or bored beyond belief all influenced its style and content. Friends, enemies, editors, and reviewers all made helpful suggestions. My wife, Jo, brought me biological aid, including glasses of wine.

Richard Cowen

HISTORY OF LIFE

ONE

THE ORIGIN OF LIFE

INTRODUCTION

Space exploration from Earth has not yet found any clear evidence of life on other planets. We still don't know whether Earth is alone in having life, but whether it is or not we must still face the question of the origin of living things. There are only three ways in which life could have appeared on Earth. It could have *evolved* from chemicals present on Earth; it could have *germinated* from some kind of spore landing on Earth from outer space; or it could have been *created* by a miracle, that is, by some event which we can never understand.

Of these ideas, the third can never be tested scientifically and therefore cannot profitably be discussed in this book. The second neatly evades the problem of the origin of life by transferring it to some other planet or to interstellar space. Fortunately, the remaining idea, that life evolved from nonliving chemicals on Earth, can be tested.

We can estimate from geological evidence the likely chemical composition of Earth's early atmosphere and oceans, and we can calculate what kinds of reactions might have gone on under those conditions. Early Earth conditions can be reconstructed in the laboratory. Chemical experiments indicate what sequence of reactions was required for the evolution of life, and how probable that sequence might have been. If we eventually find that a very improbable mixture of chemical compounds was required to combine in a set of unlikely reactions under very peculiar conditions for life to have evolved on Earth, then we'll be forced back to the drawing board to look again at other theories. But the experiments performed so far indicate that chemical evolution is a very promising way of explaining how life came to be on this planet, and they can also give us ideas about the prospect of parallel development of other life forms elsewhere in the universe.

It is generally assumed that Earth and the rest of the solar system condensed from a cloud of dust and gas about 4.6 billion years ago, though there is disagreement about details of the process. There are huge amounts of hydrogen and helium in the universe as a whole, so that Earth might have had large amounts of these gases in its first atmosphere. If so, simple reactions with carbon, nitrogen, and oxygen would have meant that Earth's first atmosphere was a mixture of hydrogen, methane, ammonia, and carbon monoxide. Many laboratory experiments on the origin of life have used this sort of mixture of gases.

But there is a fatal flaw in this argument. Both Earth and the moon are seriously short of light elements compared with the rest of the universe, particularly gases like hydrogen, helium, neon, and argon. Earth has only 1 part in 100 billion of its share of the neon in the universe, for instance. Probably the light elements were lost out into space while Earth was first accumulating from dust and gas. At that time its gravitational attraction would not be strong enough to hold in the light elements unless they were chemically bound up into larger, heavier molecules, or trapped in mineral grains. So gases like helium, argon, and neon, which do not react chemically in normal circumstances, were almost entirely lost into space, while elements like carbon, nitrogen, and oxygen were held chemically or physically in the solid particles of the forming Earth.

At any rate Earth lost its first atmosphere, but soon evolved another. Volcanoes continually erupt not only lava but vast amounts of gases, especially steam and water vapor, carbon dioxide, and nitrogen. This sort of mixture, erupting from Earth's first volcanoes, provided the atmosphere in which life first appeared. Ultraviolet radiation from the sun would have set off chemical reactions high in an atmosphere like this, producing several compounds including cyanide (HCN), which we know as a deadly poison. Cyanide is very easily dissolved, and would have been washed out of the atmosphere very quickly by rain, to accumulate in Earth's early seas. We are faced with the probability that life evolved from a mixture of gases or in an ocean that would now kill us instantly.

HOW LUCKY ARE WE?

As Earth's first volcanoes erupted steam and water vapor about 4.5 billion years ago, these gases condensed and filled up low-lying areas on the crust to form the first oceans. Water dissolves so many other chemicals (solids and gases) that it makes a very good bath in which chemical reactions can work easily. Life could probably not have evolved without liquid water. How lucky is Earth to have so much water? What are the chances of finding water (and life) on other planets?

The presence of water depends on a very delicate balance between the size of a planet and its distance from the sun. If the planet is too small, it doesn't have the gravitational attraction to hold in an atmosphere, and will lose gases, including water vapor, out into space. It will then be airless and sterile like the moon. The planet must also be at the right distance from the sun so that it receives enough heat to melt ice, but not enough to boil water.

In the solar system, all the outer planets are too far away from the sun to have liquid water. Mars is a little too small and a little too far out. It may have a little ice on its surface, but probably there is not enough for even occasional melting to support life. Venus is far too hot, being nearer to the sun than Earth. There is a good chance that its atmosphere contains clouds of sulfuric acid. If there is life on other planets, it must exist on planets circling other stars. The universe is so immense that it is practically certain that other life forms do exist. The laws of physics and chemistry are the same everywhere, and if life can originate on Earth it would be arrogant to suppose that our planet had been favored beyond the laws of probability. Remember, though, that this conclusion is only an educated guess.

STEPS IN THE EVOLUTION OF LIFE

Living things contain large and complex organic molecules, and any explanation of the chemical evolution of life must include pathways by which these molecules might have formed under Earth conditions from simple starting compounds likely to be found on the early Earth. This challenge has faced chemists since the idea of evolution among living things was generally accepted. Charles Darwin realized that his ideas on evolution by natural selection could only lead him back in time to the first living cell, and he faced squarely the problem of explaining how that first living cell came into existence. In 1871 he suggested in a letter to a friend that "in some warm little pond, with all sorts of ammonia and phosphoric salts, light, heat, electricity, etc. present, that a protein compound was chemically formed ready to undergo still more complex changes. . . ." As usual, Darwin was almost right, but it has taken 100 years to understand some of the details of the process.

Life is based on the interactions of proteins and nucleic acids. *Amino acids* are comparatively simple organic molecules that must be bound together in long chains to form proteins. *Proteins* act as building materials and as compounds helping chemical reactions in the body. *Enzymes* and some *hormones,* for example, are proteins that produce special effects in beginning and controlling chemical reactions in the body. Amino acids and proteins of various kinds are the building materials and the "operators" in the chemical factories that we call living cells. The administrative part of the chemical factory is organized in the *nucleic acids*, ribonucleic acid (RNA) or deoxyribonucleic acid (DNA), complex organic chemicals that control the formation of proteins and direct the complicated procedures by which cells reproduce. It is more difficult to make nucleic acids than proteins in the laboratory, though the steps necessary are easy to understand (Figure 1-1).

Chemists working on the problems of the origin of life generally agree on the sequence of events outlined in Figure 1-1, but there are violent disagree-

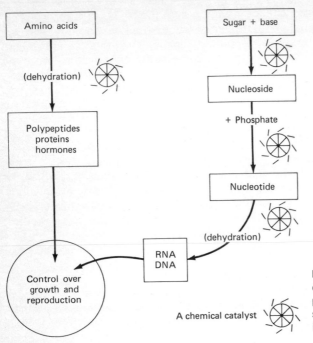

Figure 1-1 Steps in the chemical evolution that preceded life. At each step a chemical catalyst is necessary.

ments about the precise pathways involved. There is argument about the correct mixture of ingredients to be used in experiments. There must be a supply of energy to set the reactions going, and where some laboratories use electric discharge (simulating lightning on the primitive Earth), others use heat (for volcanic activity), shock waves (for meteorite impact), bombardment of high-energy particles from a cyclotron (for cosmic rays), or argon lamps (for ultraviolet radiation from the sun) (Figure 1-2). There is argument about the site of the reactions—in water, in gases, in amino acids dissolved in other amino acids, on the surface of particles, or even in outer space.

Everyone agrees on the following points. *Energy is required* to form complex organic molecules. This problem may in some cases be eased if the appropriate inorganic or organic *catalyst* is present *to help the reaction.* But continued energy input will soon break down complex organic molecules, so after they are made they must somehow *be protected from further radiation* until the next step in the process is favored. Usually, concentrations of complex reaction products are very low, and there must be some *mechanism for concentrating* them so that the next step can occur.

The account given here is a mixture of the work of different scientists, and is probably not an account of the steps which really occurred in Earth's early history. But research is active in this field, and a very important breakthrough could come at any time.

In 1953 Stanley Miller, then a graduate student at the University of Chicago, made an experiment in which electricity was passed through a methane-ammonia-hydrogen-water mixture. Several amino acids were found

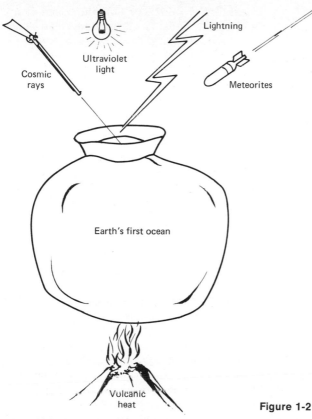

Figure 1-2 Making organic soup.

in the products. This was in many ways rather surprising, because amino acids are not very simple compounds. But it was very encouraging for those who believed in chemical evolution, and it led to many other experiments that have shown that most natural amino acids could have formed on the early Earth.

The combination of amino acids into proteins involves the loss of water molecules, and so some dehydrating agent is necessary for this next step. Sidney Fox and his research group at the University of Florida have used high temperatures for this, and have formed chemicals like proteins from dry, warm mixtures of amino acids. They believe that volcanic activity could have provided high temperatures to form proteins on the early Earth, even if only locally and temporarily.

Adenine, a base for a nucleic acid, was made chemically in 1960 from mixtures containing cyanide; this underlines the critical significance of cyanide, since adenine has a structure formed by rearranging five HCN molecules. Other bases for nucleic acids have been formed in experiments, but under conditions which would have been very rare on the early Earth.

Sugars have been formed in the laboratory by shaking a mixture of lime and formaldehyde, and other sugars were produced in conditions simulating the flow of water from hot springs over clay beds. The linkage between a base and a sugar and a phosphate to form a nucleic acid molecule (see Figure 1-1) is also a dehydration process and would have been helped if phosphates were present as catalysts. This process could have been speeded up by taking place on the surface of apatite crystals, which are rare but naturally occurring phosphate minerals.

At this stage in research, it is probably as much as anyone can do to show that the necessary reactions could have occurred *somewhere* on the early Earth. The precise conditions and sites will eventually become better known after more research. Once the process had reached the point where proteins were present on Earth, all the later steps leading to the origin of life were probably aided by proteins acting as catalysts or enzymes. The later stages in the evolution of life from chemicals probably took place not in open water, or in gas mixtures, or on the surface of crystals, but in naturally occurring concentrations of chemicals of different sorts; that is, in small "reaction tanks" where molecules would be crowded together and chemical reactions would be more rapid and more complete.

NATURALLY OCCURRING CONCENTRATION MECHANISMS

Five naturally occurring concentration mechanisms may have played a part in the later stages of the evolution of life from chemicals. *Evaporation* is one of the easiest ways to concentrate chemicals, and it is usually associated with warmth, which speeds up chemical reactions. On the Earth's surface at present, chemicals are concentrated in large lakes in hot, dry areas (the Dead Sea is the best example) and in tide pools along shorelines. Natural salt pans are really quite common features, and some important reactions could have been encouraged in these conditions on the early Earth.

Concentration by freezing has been suggested by Leslie Orgel of the Salk Institute, San Diego. He reminds us that a very alcoholic and tasty applejack can be made by freezing a mild cider. The water content freezes and leaves the alcohol and the flavoring chemicals as liquid. Although chemical reactions would be slow at freezing temperatures, they would be encouraged by high concentrations.

Very important reactions might occur in tiny droplets of chemicals as long as they contained the right ingredients. *Proteinoid microspheres* are formed when mixtures of amino acids are dried out and then mixed with hot water. Tiny spherical globules about 2μ (or 0.002mm) in diameter form spontaneously. They contain various protein mixtures, and seem to have a double-layered outer coating rather like a cell wall. Some internal reactions can take place during the formation of microspheres, as the starting mixture of amino acids is not always maintained during an experiment. Sidney Fox believes that the formation of proteinoid microspheres in hot, shallow lagoons, or around the edges of volcanic islands, may be extremely important in the development of large protein molecules.

Lipoprotein vesicles are droplets formed on a scummy surface by water turbulence or wind action. Depending on which substances might be present in the surface scum, the resulting droplets might have acted as tiny chemical reaction chambers.

Coacervate drops are formed in some colloidal solutions. Microscopic droplets rich in concentrated colloids may separate spontaneously from the solution, acting to concentrate the chemicals as much as 10,000 times. A. I. Oparin, the great Russian scientist who worked for fifty years on the origin of life, has shown that coacervate drops made in the laboratory to contain different enzyme systems will shrink, grow, coalesce, or divide much like living cells. Internally, coacervate drops can perform reactions like photosynthesis if they are provided with the right substances. Oparin believes that some coacervate drops in the early oceans of Earth could have evolved into primitive living things if they happened, by chance, to have exactly the right internal mixture of organic molecules and catalysts. Those few coacervate drops out of billions that happened to have an internally consistent mixture of proteins and nucleic acids would have been favored because they could best have absorbed and integrated nutrient compounds from the surrounding water, and could have grown and divided faster than the other coacervates around them.

THE FIRST LIVING CELLS

Living cells divide in a very precise way, under the control of their nucleic acids. Coacervate drops divide rather haphazardly, so that there is no guarantee that the resultant "daughter" drops have the same chemical makeup as the "parent" drop. Successful drops with exactly the right chemistry, growing faster than their neighbors, would tend to divide more frequently. If, by chance, among billions of drops there happened to be a set of nucleic acids that controlled the division of one drop quite rigidly, there would be some guarantee that its "daughter" drops would have that same fortunate balance of substances which had enabled the parent to grow successfully. In other words, the coacervate drop would have acquired the ability to replicate itself more or less exactly, and could be called a "living" cell. Obviously, on Oparin's model, or in fact on *any* evolutionary model for the origin of life, it would have been very difficult to draw the boundary between life and nonlife in the oceans of the early Earth, because the change was gradual.

It is not so hard to draw this boundary now. Leslie Orgel has suggested the code word CITROENS to help define living things. CITROENS are Complex Information-Transforming Reproducing Objects that Evolve by Natural Selection. All living things grow by using the information coded into the nucleic acids they inherited from their parent or parents, so that they have to be complex and information-transforming. The nucleic acids carry a code for the right combination and sequence of proteins which are to be formed, so that the growing organism is provided with a sound biochemical system. But the growing organism lives in a world that may or may not favor its growth and success. If a growing organism is "unfit," it may not survive to reproduce, or it may have fewer successful offspring than a competing organism. Thus all living things are weeded out by a continuous process of evolution by natural selection, which has been operating on Earth now for millions and probably billions of generations. As a result, the gap between life and nonlife on Earth now is very clear-cut.

So that they can grow and reproduce, living things operate energetic chemical processes aided by enzymes; thus, life causes an energy flow that we call *metabolism.* All living things therefore need energy, and they gain it in various ways.

The energy source for the growth of coacervate drops, and for the first living organisms, must have come from simply absorbing organic molecules from seawater. Most of these molecules would have been formed by the sun's radiation, and would have been amino acids, sugars, other hydrocarbons, and mineral salts. They must have formed a rich "organic soup," the accumulation of many thousands of years of synthesis. A reasonable guess is that a million years would have produced a vast amount of organic accumulation. Leslie Orgel calculates that the organic soup might have been about as nutritious as a modern packaged chicken bouillon, diluted about three times. However, coacervate drops and the first living cells must have possessed the right enzymes to attract, absorb, and break down the organic soup, releasing energy for growth and replication.

THE FIRST ENERGY CRISIS
As the first living cells absorbed the organic soup and grew and multiplied, some of them must have died. Their bodies would have provided a concentrated reservoir of organic substances to be absorbed and broken down in turn by other cells. Thus the first *ecosystem* must have developed, with some cells absorbing organic soup, and others absorbing and decomposing molecules from dead cells and from coacervate drops, in a sort of endless cycle. Eventually, however, the first living cells must have faced the first global energy crisis, as the original organic soup was used up, and as their consumption exceeded the steady but small supply of organic molecules generated by solar radiation. At this point, the biology of Earth underwent its first great irreversible change: life could no longer evolve from nonlife because the reservoir of organic soup in which large molecules could be formed spontaneously had disappeared. In the future, all formation of large molecules would take place biochemically, inside cells, aided by enzymes.

In a world with rapidly dwindling reserves of organic soup, there would have been a tremendous advantage for a cell which could absorb much of its energy directly from solar radiation, rather than from organic soup. Some simple compounds have the property of absorbing light when they are combined with a metal atom such as magnesium (as in chlorophyll). The light energy thus trapped can be used to power metabolic reactions, and in particular can be used to build up organic compounds which can be stored and later broken down to release energy.

An early cell that by chance happened to contain the right compound for trapping light energy (easily formed from simple chemicals available on the early Earth) might have been able to grow and divide more quickly than its competitors. The process by which one particular individual reproduces more quickly than others is called *natural selection,* and ensures that an individual that is particularly suited to its environment is perpetuated, whereas an individual that is not so well suited reproduces more slowly or not at all. We have seen this process acting already on coacervate drops, in prebiological or "chemical evolution." It is natural selection which acts to interconnect the individual organism with the environment in which it lives.

Light-trapping cells came to gain more and more energy as successive tiny modifications to the process led to further and more complex reactions. Finally *photosynthesis* was perfected, with the aid of the complex compound *chlorophyll.* A photosynthetic organism gains most or all of its energy from the light-triggered reaction, freeing it from dependence on outside nutrients. Such an organism is *autotrophic;* that is, its nutrition is generated internally from light energy, with only water, carbon dioxide, and a few necessary minerals absorbed from outside.

The success of autotrophic cells would have been in turn a great bonus for other cells. Nutrient organic substances would have been produced by photosynthesis much faster than ultraviolet radiation could produce organic soup. As autotrophic cells died, their bodies would have provided food parcels for other cells that had been starving on rapidly thinning organic soup. Other things being equal, the evolution of photosynthesis would have increased greatly the Earth's biological energy budget and its population of microorganisms, and should therefore have increased greatly the chances of a geologist finding one of them in the fossil record. So we must now turn to the fossil record for the real facts, instead of relying heavily on these reasonable but hypothetical lines of argument about Earth's first living organisms.

SUMMARY

Life probably evolved from chemicals on Earth soon after Earth formed 4.6 billion years ago. Earth's atmosphere would then have contained water vapor, carbon dioxide, and nitrogen, and the sun's radiation would have encouraged other compounds like cyanide to form.

Laboratory evidence suggests that the evolution of life from chemicals would have required energy and the right catalysts to help the reactions. Reaction products in the form of complex chemicals would then have to be protected and concentrated before they could combine into even more complex compounds.

Very different ideas about the actual origin of life have been suggested. All agree that conditions on the surface of the early Earth could have produced amino acids and proteins, nucleic acids, "proto-cells," and finally living cells.

The first life must have depended on chemicals for food, but later the evolution of photosynthesis allowed food to be generated more freely, using the energy of the sun's light radiation.

TWO

THE FOSSIL RECORD OF EARTH'S EARLIEST LIFE

INTRODUCTION

A fossil is a trace of an organism preserved in rock. Usually it is a hard part of an animal or plant, like a shell, a bone, or a tree trunk, more or less unchanged. Sometimes even hard parts can be chemically altered during burial, or they may be dissolved away, leaving only an impression in the rock. Occasionally, a fossil can be a complete animal, like a fly preserved in amber (fossil pine resin). Sometimes it is only a trace of the animal, like a footprint.

All kinds of factors may destroy fossils before they are found by the geologist. After death the soft parts of the organism decay or are eaten, and the hard parts may be scattered or broken by other animals or by winds and water currents. The remains, if any survive, may be damaged during burial under sand, mud, or gravel. They may be crushed beyond recognition by the pressure of tons of rock, or be burned by lava, or they may be torn apart by forces that can fracture or fold rock layers deep in the earth. Finally, they may be destroyed by erosion as the rock is weathered away at the Earth's surface. Unsuspecting geologists may walk by without seeing them, or without recognizing them as fossils. In fact, the chances of any organism being preserved and collected as a fossil are very slim. As a rule, only organisms *with hard parts* that were *buried immediately after death* are found as fossils, and then only rarely.

When we look at the fossils displayed in a museum, it looks as if we have a very good idea about ancient life. But most living organisms are microscopic or soft-bodied or both, and have practically no chance of being fossilized. When we look at a fossil collection, it is certainly a small and misleading sample of the organisms living at that time. A recent estimate suggests that we have found only 2 percent of all the species *with hard parts* that have ever lived, and practically none of those with soft bodies. This estimate, incidentally, has been criticized for its optimism! In summary, the fact that we can make some kind of story out of the fossil record is in itself rather remarkable. It says much for the imagination (or perhaps for the reckless folly) of paleobiologists who study ancient life.

The older the rock, the more time there has been for it, and the fossils it may once have contained, to be altered. We expect that Earth's first cells were small and soft-bodied, and so the difficulties of searching for them in rocks are enormous. Only a very exceptional set of circumstances would allow us to find a fossil at all in very ancient rocks. It is only since the early 1960s that we have really begun to find some traces of Earth's earliest life.

THE HARD PROBLEMS OF DATING ROCKS

The age of rocks is determined by measuring the amount of radioactive decay of various atoms included in minerals when the rock was formed. For example, rubidium 87 decays to strontium 87 at a rate such that half the rubidium has gone after 50 billion years. Apparently nothing can change this rate, so that careful and accurate measurement of the ratio of rubidium 87 and strontium 87 can give an accurate date for the age of the rock, provided that nothing serious has happened to the minerals in the meantime (such as remelting the crystals, or dissolving out the strontium).

Usually, in dating a rock, several different methods are used to give a more reliable result. The potassium-argon method and the uranium-lead method are similar in principle to the rubidium-strontium method. All these methods work best for igneous rocks, which were formed from molten rock—the methods then give the time (perhaps within 5 percent error) at which the mineral crystals formed. Metamorphic rocks have been altered by high temperatures or by high pressures, and here the dating methods often give the time at which the mineral crystals were altered, not the time at which they first formed. Even if we can tell the age of an igneous or metamorphic rock, it doesn't help much in telling the age of fossils, because fossils are hardly ever found in rocks like this.

Fossils are usually found in sedimentary rocks, formed under ordinary conditions on the earth's surface. Sediments very rarely have minerals in them that will tell the time by radiometric methods at which they were laid down. It is therefore very difficult to date fossil-bearing rocks in *absolute* terms, that is, in exact numbers of years. Instead, one hopes to find the rare cases where two layers of igneous rock, such as lava flows, lie above and below a layer of sediment with fossils in it. By dating the lava flows, one may be able to estimate the age of the fossils in the sediment.

It is much easier to tell whether one fossil collection is older than another than it is to tell *how much* older it is. Paleontologists work with a *relative* time scale, one which says "A is older than B" rather than "A is 3 billion years old and B is 2.72 billion years old." This will be discussed further in Chapter 6.

Carbon 14 dating is one method that can be used to date the age of organic matter (including fossils) directly in exact numbers of years. However, carbon 14 breaks down to nitrogen 14 so that half of it is gone after only about 5,730 years. The method is very useful for historians and archeologists, who work in this kind of time span. But it is useless for most geological work, because the time spans involved run into millions and billions of years.

Earth's Oldest Rocks   The age of the solar system is about 4.6 billion years, from dates measured on meteorites. This is an incredibly long time—4.6 billion is the number of seconds in 150 years, for example. The oldest known rocks on Earth are found in northwest Greenland and in Minnesota, and are about 3.8 billion years old. But these are igneous and metamorphic rocks, formed either molten or under very severe pressure and temperature. The oldest known *sediments* were found in west Greenland in 1973, and seem to be about the same age, 3.7 billion years. But even the optimist who might look for fossils in these sediments would concede that the oldest one-sixth of Earth's total history offers no hope of finding any fossils at all.

Fortunately, this may not be too tragic. It seems that the sedimentary rock record is long enough to show the emergence of life. We do not see real fossils in the earliest sediments, but we can look at indirect evidence. Carbon isotopes can be used to check for the existence of photosynthetic cells in the following way.

There are two main carbon isotopes or types of carbon atoms—carbon 12 and carbon 13. Carbon dioxide in the air tends to be richer in the lighter carbon 12 because it evaporates more easily from wet surfaces. The sea, then, is slightly enriched in the heavier carbon 13 which is left behind. Carbon 12 also enters living cells during photosynthesis more easily than does carbon 13. Some early sedimentary rocks (the Onverwacht Group in South Africa, about 3.35 billion years old) show that the carbon 12 content rises sharply through a series of rock beds, probably marking the time when photosynthesis involving carbon dioxide first began.

Earth's Oldest Fossils   We have seen that photosynthesis probably developed fairly soon after life first appeared, when the first global energy crisis overtook the organisms living on organic soup. So the origin of life on Earth was probably not long before 3.35 billion years ago. The most difficult step of all had been taken.

The rock record suggests to us, therefore, that organisms lived on Earth about 3.35 billion years ago. Fully 150 million years later (!) we see the first real fossils. Our present experience with ancient rocks shows that cells are most likely to be preserved in cherts, which are resistant sedimentary rocks formed by the precipitation of dissolved silica (SiO_2). Chert impregnates and preserves even minute structures like cell walls. Cherts are more or less watertight, so that water cannot trickle and percolate through them to dissolve or contaminate the fossils.

Early in the 1960s, Elso Barghoorn (from Harvard University) and J. William Schopf (now at the University of California, Los Angeles) discovered numerous specimens of single cells preserved in the Fig Tree Chert in South

Africa. This rock unit is approximately 3.2 billion years old and contains the earliest fossils so far discovered, a spherical blue-green alga and a bacterium. The alga is well enough preserved to show that it has a granular surface, and inside some specimens there are lumps of organic matter which may even represent rotted fragments of the cell contents (Figure 2-1).

In rocks a little younger than the Fig Tree Chert, we find traces of large mats of blue-green algae, forming rock masses called *stromatolites*. Living forms can be studied today in a few rare supersalty lagoons like Shark Bay, Western Australia, where mats of blue-green algae grow luxuriantly and photosynthesize in warm shallow water that is too salty to allow grazing marine animals like snails and sea urchins to live there and destroy them. The algal mats are made up of innumerable tiny threadlike filaments that are rather slimy and sticky. When the tide comes in at Shark Bay, it covers the algal filaments with sediment which sticks to the surface of the mat, but the algal filaments slide and grow through the sediment layer back into the light. This cycle repeats itself daily, until mounds of layered algae and trapped sediment are built up in shallow water. Fossils with exactly the same structure as these living algal mats are known from the Bulawayan Group of rocks in Rhodesia. They are at least 2.9 billion years old and are probably more like 3 billion years old. Stromatolites are important for two reasons: first, they show that blue-green algae had evolved from single spherical cells to a point where they formed filamentous mats; and second, the algal mats must have been producing important quantities of oxygen at this time (Figure 2-2).

THE OXYGEN REVOLUTION
Earth's early atmosphere contained no free oxygen, and all the organisms on the early Earth must have been anaerobic (living without oxygen). Oxygen

Figure 2-1   Earth's earliest fossils, from the Fig Tree Chert of South Africa, about 3.2 billion years old. (*a*) *Archaeosphaeroides,* an algal cell; (*b*) *Eobacterium,* a bacterial cell. (Photographs courtesy of J. William Schopf: (*a*) from Schopf and Barghoorn, *Science* 156, 508–512, 1967; (*b*) from Barghoorn and Schopf, *Science* 152, 758–763, 1966. Copyright 1967 and 1966 by the American Association for the Advancement of Science.)

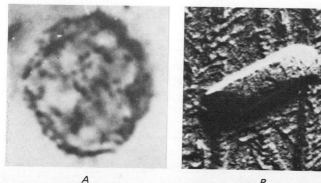

A   B

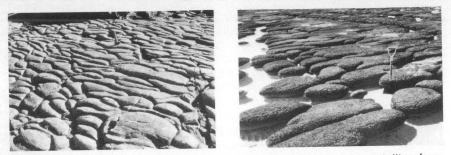

Figure 2-2 Ancient stromatolites and modern algal mats. Ancient stromatolites, from Precambrian rocks near Great Slave Lake, Canada; modern algal mats, Shark Bay, Western Australia. (Photographs courtesy of Paul Hoffman.)

first produced by photosynthesis would have been quickly taken up to oxidize iron, carbon monoxide, and any unstable organic matter in seawater. Oxygen is a very powerful poison to anaerobic organisms, and its appearance in seawater must have spelled disaster for many of Earth's first organisms. Those that survived either became limited to oxygen-free environments, or they had to develop biochemical defenses against oxygen. A long period of evolution was required to make these biochemical changes. Since oxygen was at first probably produced locally in small quantities which were quickly absorbed by other chemicals, a long time passed before it affected the whole world.

Karl Marx said that each society contains the seeds of its own destruction. This is certainly true of Earth's first organisms; by their development of photosynthesis as a reaction to their energy crisis, they revolutionized their world by "polluting" it with oxygen. Many of them must have perished in the resulting crash; some refugees like methane-producing bacteria are now confined to habitats where free oxygen is not found, such as stagnant swamps.

For organisms able to tolerate oxygen by developing biochemical resistance, there was a great opportunity to increase the energy available to them. Anaerobic cells break down molecules by fermentation, like yeasts which make wine or beer by breaking down sugar in juices. However, ten times as much energy is released by aerobic cells, which can use oxygen to "burn up" food completely in the process of oxidation. Those cells which adapted in this way to the newly oxygenated seawater were thus at a tremendous advantage in terms of energy.

None of Earth's early single-celled organisms could engulf or "eat" other cells: the largest molecules that can pass through the wall of cells like bacteria or blue-green algae are nucleic acids or proteins. So there were no predators in the early seas of Earth, but conditions were far from being a kind of paradise. There were no checks on population sizes except starvation or overcrowding. There must have been incredibly rapid fluctuations of populations, particularly among photosynthetic cells, as day and night, and summer and winter, altered their productivity. There must have been the most spectacular algal blooms (or booms!) the world has ever seen, followed by wholesale deaths as food supplies became locally exhausted. The algal blooms would have set off tremendous waves of oxygen spreading through the water, wiping out whole populations of anaerobic organisms, but rapidly diminishing as the algae died off for want of nutrients, as the oxygen was diffused away and absorbed by other chemicals, and as the organic matter decomposed.

The chemical balance of seawater would have been altered by the wholesale release of oxygen. The earliest ocean was probably rather acidic, with a high CO_2 content. Oxygen released by photosynthesis would change this and lower acidity. Modern seas are now stable and salty, not acid, so that during the oxygen revolution the chemistry of seawater probably underwent a fundamental and irreversible change.

We can see the *geological* evidence for the oxygen revolution in the deposits called *banded iron formations,* which are important rocks laid down only between 3 and 1.8 billion years ago. They are thick sequences of rock, made up of very thin bands of iron ore and chert. Chert is common in these ancient rocks because silica was present in the earliest seas in fairly large amounts; there were no organisms at that time using it to build skeletons.

In the Hamersley Range of Western Australia, the banded iron formations contain very thin bands that are only millimeters thick but can be followed for several kilometers. Similar banding exists in other iron formations in other continents. In the Gunflint Chert of the Lake Superior region, there are cherty mounds that are banded with greater and lesser amounts of iron minerals.

If we look at the banded iron formations in terms of the biology of the world during the oxygen revolution, we can see that algal blooms could have released waves of oxygen that changed seawater chemistry enough to precipitate out bands of iron minerals. In the Hamersley iron deposits, the algal blooms were probably seasonal, forming algal scums on the surface of a great stagnant salty pond. Each year would have been marked by a thin band of sediment laid down on the sea floor. In the Gunflint Chert, the banding was more local, and the algal activity probably occurred daily in small mats on a shallow sea floor rather than in a surface layer. In a local area, day and night would be sufficient to cause different rates of iron deposition if the water chemistry happened to be balanced just right.

The last banded iron formations, at about 1.8 billion years ago, mark the time when enough free oxygen had been produced to sweep the sea permanently clear of dissolved iron by combining with it and producing iron minerals. After this the seas were essentially oxygenated, although perhaps not up to modern concentrations. This would also mean that there would be free oxygen in increasing quantities in the atmosphere. The sun's radiation acts on oxygen high in the stratosphere to produce ozone, a gas which is O_3 rather than the typical oxygen molecule, which is O_2. Even a very thin layer of ozone can block off ultraviolet radiation rather effectively. Ever since the oxygen revolution, then, Earth's surface has received comparatively little ultraviolet radiation. This may sound like bad news to sun worshipers, but the fact is that modern levels of ultraviolet radiation in sunlight are about all that we and other animals can safely tolerate. The development of the ozone layer, blocking ultraviolet radiation, has meant that the production of new "organic soup" in the oceans has stopped. All organisms are now dependent ultimately on photosynthetic plants that trap the visible part of the sun's radiation, rather than the ultraviolet part.

On land, the presence of oxygen in the air for the first time would have produced rusting of any iron minerals exposed at the surface by weathering and erosion. Rivers would have run red as they flowed across the early Earth's surface after the oxygen revolution, before any vegetation evolved on land. In fact, rather than the oxygen revolution producing a "greening" of Earth, it was probably a Red Revolution. In confirmation of its timing, it was at about 2.0 to 1.8 billion years ago that the first important red-colored sediments were formed, right at the end of the banded iron formations.

Toward the end of the oxygen revolution, we have a glimpse of the early organisms involved in it. The Gunflint Chert of the Lake Superior region, about 2 billion years old, is a famous thin rock unit that extends for 150 miles. It is famous, among geologists at least, for its magnificent preservation of microscopic single-celled fossils. They are mostly algae that form stromatolitic mats, but there are many other microorganisms, including blue-green algae, bacteria, and a mysterious fossil called *Kakabekia* that is not understood at all.

SUMMARY

Fossils are only a remnant of ancient life, and a lot of guesswork is involved in reconstructing ancient biology. Rocks are often difficult to date, adding further problems. The little evidence we have about Earth's earliest life suggests that life and photosynthesis appeared a little before 3.35 billion years ago. The first fossil cells are about 3.2 billion years old. Algal cells produced oxygen by photosynthesis, and in so doing they poisoned the environment for many of the early organisms. By about 1.8 billion years ago, only cells that could tolerate or use oxygen were widespread on Earth. The chemical changes caused by the oxygen revolution can be seen in the rock record as well as in the fossil record.

THREE

THE EVOLUTION OF SEX

CELLS WITH NUCLEI

All the organisms in the Gunflint Chert are *procaryotic,* that is, they did not have a cell nucleus, and the division of the cell into two offspring did not include a precise duplication of the genetic material (the DNA). Only molecules could pass through the cell wall. But procaryotes were and are highly successful. On the early Earth, in an unstable and highly variable world, cells with wide variability and very rapid reproduction would have been favored, so as to meet sudden changes in the environment. The cells would have been rather simple and comparatively tough.

The oxygen revolution greatly stabilized the environment, and this would have favored the evolution of organisms which were less variable and were more precisely adapted to a given environment. For the first time, genetic stability rather than genetic flexibility became important. Organisms could then become more complicated in their structure and their biochemistry. In the same way, frontier conditions in the Old West produced (according to legend, anyway) individuals who were very tough and adaptable, each capable of doing a rough but adequate job of hunting, mining, house building, first aid, and dealing with emergencies. In today's more stable conditions, social evolution has favored the individual who lives in one community and does only one thing extremely well, like repairing telephones or teaching geology classes.

Eucaryotic cells represent these modern, more specialized members of society. In contrast to procaryotic cells, eucaryotes have their DNA organized into specific packets (called *chromosomes*), which are in turn packaged up into one parcel (the *nucleus* of the cell). The cell reproduces in an exact way so that each chromosome is duplicated precisely. Thus a successful eucaryotic cell normally divides into two identical "daughter" cells, which will grow up looking exactly like their parent and will be equally successful unless the environment has changed. Eucaryotic cells, with their exact genetic duplication, are obviously well fitted for comparatively stable environments—requirements for success won't change much in one generation, and what's good for the parent will also be good for the offspring. Occasionally there may be a genetic copying error, so that a cell has a changed DNA content—it is a *mutant* cell. Usually mutants are very badly adapted for life, but very occasionally they are better adapted than their parents. The occasional mutation provides enough novelty in the population that some changes in the environment can be handled successfully.

Eucaryotes can be divided into plants and animals—although one-celled organisms are sometimes hard to classify. Plant cells have a rigid coat of cellulose which protects them but also prevents them from moving freely. They cannot travel to gather food—instead, they use photosynthesis to build up nutrient compounds like sugars. Animal cells have elastic walls and can often "creep" after food; animals usually search for plants or other animals for food rather than manufacture their own. These basic differences between animals and plants date back to the Precambrian times when the world was still populated only by single cells.

Simple eucaryotic animal cells can engulf large particles and can excrete large particles through the cell wall. In other words, they can eat food instead of just absorbing it. They do this by using the elastic cell wall which also allows them to "creep"along. These adaptations for moving and feeding are obviously great advances over the procaryotes, and are fundamental steps in the evolution of higher living animals.

The procaryotic ancestor of eucaryotic animal cells was probably an absorbing cell depending on organic molecules for food. The ability to eat large organic fragments would have been gradually improved until an efficient scavenging eucaryote was developed, capable of absorbing quite large pieces of dead tissue and rotting debris. Finally the eucaryote would evolve until it could eat and digest entire cells, and it would then be the first predator eating living organisms.

The evolution of predators was a very important ecological step. Predators control population bursts in their prey by eating off the surplus, and so the *giant* algal blooms of earlier times could not occur any more after eucaryotes appeared. In addition, predators tend to encourage a greater variety among their prey. If one kind of prey becomes very numerous, the predators enthusiastically eat it back; thus one prey organism finds it difficult to expand to the point where it totally outcompetes another. So we should expect that with the appearance of eucaryotic predators the microorganisms of the world would become more varied. There is not enough fossil evidence yet collected to test this idea.

Eucaryotes (including people) "burn" their food internally using oxygen, and so they need an oxygen-rich environment. Eucaryotes cannot have evolved before the oxygen revolution was largely complete at about 1.8 billion years ago.

The only way of detecting the evolution of eucaryotes in the fossil record would be to find a fossil cell with a preserved nucleus. This is obviously difficult, since the nucleus is a soft structure which is hardly ever preserved even in the most recent fossils. J. William Schopf has found cells with internally preserved spots that look suspiciously like nuclei in some chert beds from Australia, ranging from 0.9 to 1 billion years old (Figure 3-1). There are less convincing spots in cells from the Beck Springs Dolomite

Figure 3-1 Eucaryotic cells from the Bitter Springs Chert of Australia, 900 million years old. Photographs (a) to show "spots" and "spots with membranes" which are probably remnants of nuclear and cytoplasmic tissues. Four cells (b) are shown which died in different stages of division. (Photographs courtesy of J. William Schopf.)

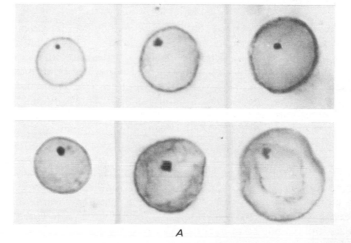

A

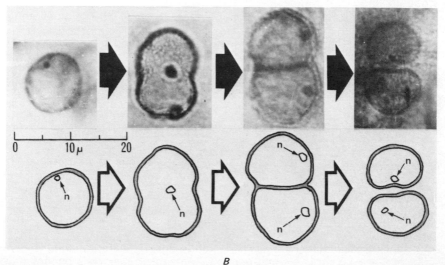

B

of California, about 1.3 billion years old, and there are really dubious but possible eucaryotic cells from the Belcher Group of rocks from Hudson's Bay, Canada, at about 1.7 billion years. This gives some idea of the far-flung search for evidence about the appearance of eucaryotic cells, but even more evidence is desperately needed.

THE EVOLUTION OF SEX

Modern eucaryotes do not necessarily reproduce simply by splitting, but have sexual reproduction. This is a mechanism for "shuffling" together the DNA of two individuals, and "redealing" it to their offspring, which are therefore similar to, but not identical with, their parents. In fact, modern genetic studies show that there is an impossibly low chance of two sexually reproduced individuals being genetically identical, unless they developed from the same egg (like identical twins).

Eucaryotes are so complex that only two closely similar cells can shuffle their DNA together with any chance of producing a completely operational set of offspring. In eucaryotes, therefore, there are complicated mechanisms (physical, behavioral, "instinctive") to ensure that only genetically similar individuals will engage in sexual reproduction. We call such a set of individuals a *species,* defined genetically as a set of individuals that shares the same *gene pool* (quantity of DNA). Only in sexually reproducing organisms does this definition of a species have any real meaning.

Sexual reproduction results in offspring that resemble their parents in all major characters, but are unique in their particular combination of minor characters. The process produces a great deal of variability: some offspring are slightly better fitted to the environment than their parents, some are slightly worse, and most are about the same. If the environment is likely to change at a rate that is slow compared with the life-span of one generation, then there will usually be some offspring well suited for the slowly changing conditions.

In a nonsexually reproducing population, the only way for a favorable mutation to spread through the group is for its original possessor to outdivide the others. The environment selects or rejects the whole DNA package of the individual with the new mutation, and this individual either succeeds or fails. This one-shot chance has two bad effects. Many favorable new mutations can be lost because they originate in individuals with other poor characters. Conversely, a favorable mutation might allow one individual to reproduce innumerable replicas of itself, so that the whole population became genetically alike. Such a population could then be wiped out by a sudden change in the environment.

In a sexually reproducing population, a mutation is shuffled into a different combination of genes in each offspring of the mutant. Natural selection can then operate on many different "test individuals," each having the new mutation combined with a different set of genes. Those *combinations* which are favorable can be passed on, and those which are disastrous will be unfit and the individuals with them will not survive. So a sexually reproducing population not only tends to have more genetic variability than a nonsexually reproducing one, but it can also respond more smoothly to changing environments. In favorable circumstances, evolution can be greatly accelerated this way.

Figure 3-2 Evidence of sexuality in cells from the Bitter Springs Chert. (*a*) The fungus *Eomycetopsis*; (*b*) an alga with cell tetrads. *Eotetrahedrion*. (Photographs courtesy of J. William Schopf.)

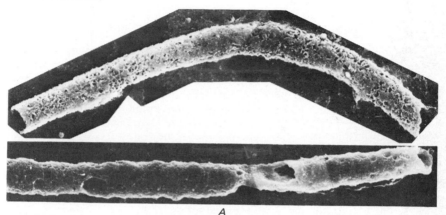

A

B

Yet there is a built-in conservatism in sexual reproduction. Particularly favorable mutations are diluted in the recombination or "shuffling" of genes. Children are not replicas of parents, even though they may carry much the same set of genes; they are unique individuals. Unfavorable genes, too, may be held in the gene pool if they occur in combination with other genes that compensate for them. The result is that populations are not made up of numerous replicas of Leonardo da Vinci, Beethoven, or Shakespeare (or Attila the Hun or Genghis Khan), but consist largely of so-called "average" individuals, some of whom may be carrying fragments of the gene set of outstanding ancestors.

Since sexual reproduction has so many advantages, it is very important to get some idea of when it was first invented by eucaryotes. There are two methods which can be used: first, to identify fossils of organisms that definitely reproduced sexually, and second, to identify some special sexual structure. J. William Schopf has found both kinds of evidence among eucaryotic cells from the Bitter Springs Chert of central Australia, about 900 million years old. The chert contains fossils that are probably fungi, and therefore probably reproduced sexually; and it also contains algal cells arranged in a group of four (Figure 3-2). This usually, but not always, indicates that a cell division (meiosis) has occurred that is a definite part of sexual rather than nonsexual reproduction. Thus we can be sure that the eucaryotes had developed sexual reproduction by 900 million years ago. We still don't know how long *before* 900 million years ago sex was invented. But it is rather poetic, if rather gloomy, to think that it might have been at a place called Bitter Springs.

SUMMARY

Cells with nuclei evolved at the end of the oxygen revolution. Simple animal cells could then eat other living cells for the first time. Sexual reproduction evolved. Although this is a much more complicated process than simple cell division, it has many advantages. The timing of these evolutionary events is not yet well known, but cells with nuclei and sexual reproduction had been evolved by 900 million years ago.

FOUR

HOW TO BUILD ANIMALS

SOME THEORY

Because sexual reproduction allows organisms to combine variability with complexity and strict genetic organization, it is not surprising that after the time of the Bitter Springs Chert, evolution seems to have been rapid. Between 900 and 750 to 700 million years ago, the single-celled organisms seen at Bitter Springs evolved into animals that were large and complicated and had special organs and tissues.

We have no direct fossil evidence of the origin of complex multicellular animals (*metazoans*) because they were fully developed by the time we first see them as fossils. We can only look at the tremendous variety of living animals and try to reason out what the early stages of their evolutionary story might have been.

There are three simple kinds of multicellular animals: sponges, coelenterates, and flatworms. A sponge is essentially an organized and supported mass of flagellate protozoans. A flagellate protozoan is a single cell carrying a lashing filament that moves it through water, rather like the single oar of a primitive canoe. A mass of such cells, organized into a sponge, does not move the sponge but forces water through the sponge. In the process the cells filter bacteria from the water. The other cells which go to make up the architecture of a sponge serve as digestive cells, as cells which make up a kind of skin, and as cells to form a simple skeleton made up of many fine needles of silica or calcite, or protein fibers.

Coelenterates include corals, sea anemones, and jellyfish, and are a little more complex: they have a two-layered structure, with an outer skin and an interior layer of cells that are mostly digestive. Sea anemones and corals have the two-layered tissue arranged to form a cup surrounded along its edge by tentacles. Prey is seized by tentacles or trapped in sticky mucus, and is then carried inside the cup where the digestive cells break it down with powerful chemicals and absorb it.

The third kind of simple animal, and the one that probably gave rise to all the other metazoans, is a flatworm, which differs from the other two in having considerable powers of movement and a much better coordination between cells. A quite well developed nervous system acts to coordinate the activity of muscles so that a flatworm can quickly and efficiently react in a fairly mobile and sophisticated way to external stimuli. Flatworms prefer to move in one direction, and have a primitive head at the front. They move along the bottom by creeping on a broad, flat foot.

These simple animals mostly live on the sea floor, and while the sponges and most coelenterates are attached to the sea floor and depend largely on food particles in the water, the flatworms are mobile scavengers and predators on the sea floor.

More highly developed metazoans are wormlike animals of various kinds, with a special fluid-filled internal cavity, the *coelom.* Almost all higher animals are coelomate (in man, the coelom is the sac containing all the internal organs). The similarity of coelomates suggests that they evolved from one ancestor, probably wormlike, which in turn developed from a flatworm by forming the coelom. This must have been evolved for some good reason.

An English zoologist at the University of Newcastle, R. B. Clark, has suggested that the invention of the coelom allowed animals to make a great evolutionary breakthrough. Liquid is incompressible, and a flatworm that first developed any kind of internal fluid pool would have been able to squeeze it by body muscles, automatically poking out the end of the body. This could then have been used as a kind of power drill to allow the worm to burrow into soft sediment. There would have been abundant food resources in rich organic muds, and the new evolution of the coelom would have allowed animals to reach this food resource below the sediment surface for the first time. In addition, burrowing worms would have been able to escape from flatworm predators on the surface. Clark believes that many different early flatworms took up a burrowing way of life for these reasons.

The coelom would have provided another great advantage for early metazoans. Animals rely on respiration, which means that a supply of oxygen must reach all cells. Single-celled organisms can get all the oxygen they need by diffusion from the water, but larger animals cannot pass enough oxygen through thick tissues in this way. The coelom, if it had some oxygen-carrying molecule such as hemoglobin, would have allowed larger and more complicated tissues, organs, and animals to develop by carrying oxygen in its internal fluid and thus acting as a simple *circulatory system.*

If Clark is correct, respiration problems would have been particularly serious for early coelomates because they were burrowing in rich organic muds, which are very low in oxygen. Therefore we would anticipate that some special organs might be developed to obtain oxygen from the overlying seawater while the main body of the worm remained below the surface for protection. Thus we find that many coelomates that do not burrow very actively have filaments and tentacles and gills of various sorts as specialized *respiratory organs* extending into the seawater. It is a very short step from here to a point where the coelomate actually collects food as well as oxygen from the seawater (as in the bryozoans, brachiopods, some molluscs, some echinoderms, some worms, and early chordates). Other worms seem to have been able to burrow actively enough that their movements inside their burrows brought enough oxygenated water flowing over them. In these groups, respiration through the skin was sufficient for their needs as long as an efficient internal circulatory system was developed (as in some worms, arthropods, echiuroids, and so on).

SOME FACTS

This evolutionary model is speculative, but fits with the very poor fossil record of Precambrian animals. The best evidence comes from Australia, where we find a record of deposits laid down between 800 and 650 million years ago in a broad shallow sea stretching widely over the continent.

At about 800 million years ago, in the Buckingham Sandstone of northern Australia, there are fossil burrows set vertically in the sediment, showing that the animals that made them were rather sluggish, probably wormlike suspension feeders. However, after this glimpse of earliest worm-like animals, the record is interrupted by a great ice age, which may have lasted for about 100 million years. It looks as if this early ice age was also very extensive in area, because there is evidence for it on other continents too.

At the end of the ice age, there is much more evidence that metazoans had become well-developed animals. A variety of tracks and trails has been preserved in the rocks, and in very favorable circumstances impressions of soft-bodied animals were preserved. The same sets of traces have been found as fossils on several continents, so that the sea floor of the world was then probably populated by a group of animals known as the *Ediacara fauna* (after the Ediacara Hills in South Australia where many specimens have been found).

The Ediacara fauna lived in shallow water in sands and muds, and floating in the water, from about 700 to 600 million years ago. Twenty years of collecting in the Ediacara Hills have produced over 1,400 specimens of body impressions, and many more tracks and trails. Worms of various kinds patrolled the sea floor, some moving by squirming through the sediment, and others moving largely by "walking" on bristle brushes located on their body segments. Fossil burrows in the sediment show that some worms were a centimeter across, and were deposit feeders leaving trails of fecal pellets behind them. Smaller worms left trails on the surface as they wriggled through the top layer of sediment. Many jellyfish impressions show that these animals were floating in the water above the sediment. Smelly experiments done with dead modern jellyfish show that the Ediacara animals must have been covered over with sediment within a few days after death in order to preserve even the impression of the body. The other important members of the Ediacara fauna are colonies of coelenterates called sea pens, which superficially resemble plants, and are adapted like corals to capture and eat floating animals in the water column. There are several fossils at Ediacara that are not remotely like any living animals, and research on these is continuing (Figure 4-1).

At Ediacara we have our only good view of early soft-bodied metazoans. There were some who fed on floating animals, but most were mud eaters (deposit feeders). We do not see any evidence of animals living on algae on the sea floor, or of suspension feeders filtering the water. Probably the Ediacara fauna was only one specialized or unbalanced community which happened to be living in conditions favorable for its preservation in the fossil record, while other early animal communities have not been preserved.

THE EVOLUTION OF SKELETONS

One of the most spectacular events in the history of life was the development of hard parts. Numerous new animal groups with skeletons appeared in the fossil record for the first time about 600 million years ago. The event is

A

B

C

Figure 4-1 Soft-bodied fossils from the late Precambrian Ediacara beds of South Australia. (a) *Cyclomedusa*, a jellyfish; (b) *Dickinsonia*, probably a worm; and (c) *Tribrachidium*, unlike any other known organism. (Photographs courtesy of Professor M. F. Glaessner.)

marked by the recognition of a new era of geological time—the *Paleozoic*—to succeed the Precambrian, which is practically barren of fossils. One would imagine that the task of the paleontologist would become easier because of the numerous Cambrian fossils (570 to 500 million years ago) available for study, but in fact the appearance of hard parts poses a difficult problem.

First, several major groups of animals appeared almost at the same time in the fossil record, although they were very different from one another (Figure 4-2). Conventionally, animals are grouped into a *phylum* when they are obviously related. At the same time, each phylum has its own particular body structure and its own particular kind of ecology. Thus Mollusca or Chordata are phyla, reflecting the fact that there are no intermediate animals, dead or alive, linking these major groups of animals with any others. So in the Early Cambrian period, when no less than six major phyla appeared in the fossil record for the first time, it is obvious that there had been prior evolution in the Precambrian during which their ancestors evolved and separated from other animals. We do not know whether this Precambrian evolution was slow or rapid, but it must have taken place in soft-bodied ancestors that are not preserved in the Ediacara fauna. We have to make up a reasonable story for a

Figure 4-2   Diagram to show the range of some important fossil groups.

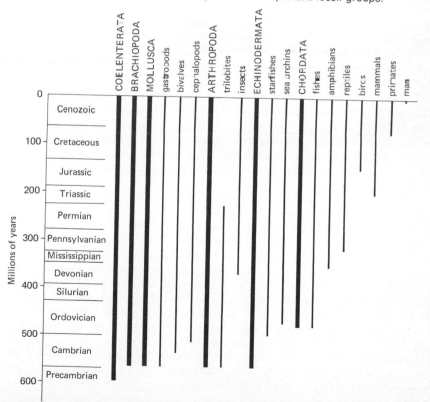

sequence of evolutionary events that would produce the separate phyla by the time we see them with hard parts in the Early Cambrian. Each phylum, operating its different biological way of life, must have evolved by a different evolutionary pathway.

In trying to work out the time of appearance of each phylum of animals, we must take into account how much the development of a skeleton would have affected an animal. A sponge is much the same whether or not it has an internal mesh of mineral crystals, or a soft protein like a bath sponge; but a clam without a shell is difficult to imagine. Therefore, the phylum Mollusca (which includes clams) probably did not really exist before its ancestors developed shells in the early Cambrian: it evolved from something else.

Thus there is a problem in the *variety* of phyla that suddenly became visible in the Early Cambrian fossil record. Another problem is that the various animals developed hard parts, of different structure and of different chemistry, *at about the same time.* Sponges developed a fine meshwork of tiny silica needlelike crystals interwoven with the cells making up their tissue. Brachiopods and arthropods developed a strong external shield over their soft parts. Different brachiopods used calcium phosphate or calcium carbonate, and arthropods used a protein called chitin, impregnated in different groups by calcium phosphate or calcium carbonate. Echinoderms developed a flexible *internal* skeleton just under their skin, made up of small separate plates of calcium carbonate. Molluscs secreted *external* calcium carbonate shells.

The reason behind the development of skeletons is not, for example, that calcium suddenly became available in the ocean. Hard parts were constructed in so many different animals by so many different biochemical methods that we can be sure we are looking at an important event in the evolution of the world's life, arising from ecological rather than chemical reasons. What were these reasons?

For some animals, the skeleton gives support for tissues, allowing a stronger, larger construction. Thus sponges would have been able to grow larger and more erect, extending further into the water column so that they could reach more food.

For some animals, the skeleton would have provided a hard "box" inside which the vital body organs could work in a more controlled microenvironment. Physical protection is part of this effect, but in the absence of obvious large Cambrian predators, it is not clear that this would have been the main reason. But general internal stability is an advantage for any organism. Early brachiopods would have been able to filter water inside an internal shell cavity, away from outside currents and eddies, an advantage not available to their soft-bodied ancestors, which had no shells.

For some animals, hard parts may have served a special function of their own, opening up a completely new way of life. For trilobites, for example, Richard A. Fortey of the British Museum of Natural History has suggested that their evolution of a large head shield may have been a vital step in burrowing technique, shoveling sediment aside like a bulldozer instead of squirming through it like a worm.

Altogether, the common factor in the evolution of skeletons in the Early Cambrian is the dramatic invasion of new ways of life on a shallow sea floor.

Sponges may have been involved in filtering bacteria; trilobites would have been digging powerfully in the sediment; molluscs were creeping on and just in the sediment surface; and a few brachiopods and echinoderms were feeding from the water, slightly above the sediment surface. All these ways of life would have been impossible, or would have been less efficient, without hard parts. We are left with the impression that the world had rather quickly become favorable for this rapid evolution of new ways of life. Why was this? The answer will be explained after we look at more data.

EARLY PALEOZOIC LIFE IN THE SEA

The Early Cambrian was a rather dramatic time, but other important events occurred later in the Cambrian. In general, they consisted of further exploitation of ways of life in the sea, filling out major groups of animals from their initial pioneer forms, in much the same way that a sophisticated society soon followed pioneers into the early West. It is fair to say that most groups of Cambrian animals were deposit feeders, eating mud, in particular the trilobites that dominate most Cambrian fossil faunas.

There is an exciting sidelight on this picture. In British Columbia there is a thin bed of shale high on a mountainside, the Burgess Shale of Middle Cambrian age. Here in 1909 Charles D. Walcott of the Smithsonian Institution discovered an amazing variety of fossils of soft-bodied animals, preserved so perfectly that he was able to work out tiny details of legs and antennae on the arthropods, and to see features like guts and intestines. The importance of the Burgess Shale fauna is that it contains an overwhelming proportion of deposit feeders, just like the "normal" Cambrian hard-part faunas. So we can be satisfied that we can (sometimes at least) draw reasonable conclusions from our examination of hard-part fossils in dealing with the paleoecology of ancient environments (Figure 4-3).

Another major change in world faunas took place about 500 million years ago at the beginning of the next geological period, the Ordovician. It was not as dramatic as the early Cambrian event because it was more gradual. Essentially, there was a tremendous upsurge in the variety and number of suspension-feeding animals, taking food from the water. Thus brachiopods, bryozoans, graptolites, bivalve molluscs, and suspension-feeding echinoderms all went through explosive evolution, while trilobites suffered some rather important extinctions. Certainly, deposit-feeding animals like trilobites were still very successful and numerous in the Ordovician, but their *dominance* of marine faunas was broken forever.

Just as before, we can make up a list of advantages for the new suspension-feeding animals. Bryozoans and echinoderms formed skeletons that enabled them to stand erect in the water currents to filter water effectively. Graptolites were colonial animals, related to chordates, that floated in the water and presumably filtered food from it. But these details hide the overall conclusion that in Early Ordovician times many different animals were beginning to depend successfully on floating plankton as a food resource, whereas other animals that continued to rely on food from sea-floor sediments were not so successful.

The development of these new animals in the ocean was accompanied

A   *B*

Figure 4-3 Soft-bodied fossils from the Burgess Shale, Middle Cambrian, Canada. (*a*) *Naraoia*; (*b*) *Aysheia*; both are related to arthropods but their true nature is uncertain. (Photographs courtesy of the U.S. National Museum.)

by yet another new way of life. Two different groups, the starfish and the cephalopod molluscs, both of which appeared in numbers in the Early Ordovician, evolved a way of life that was at first one of wandering opportunism, preying on anything dead or alive that they could eat. Some later specialized to take on a mainly carnivorous way of life. Most fossil cephalopods look like the living pearly nautilus, with beautiful chambered shells. There may have been cephalopods without skeletons like most living ones (octopus and squid) in the past, but if so they have left practically no fossil record. The chambered shell of the living nautilus and of most fossil cephalopods is fascinating because it makes them like submarines with buoyancy tanks, giving them precise control over balance and buoyancy in the water. Cephalopods were the first animals to solve the many problems of a free-swimming life in open water, and they quickly became very diverse and numerous in Ordovician seas. One species grew to about 30 feet long!

There is an ecological principle that predators tend to cause diversity among their prey. If one kind of prey becomes too numerous, it is preferentially eaten until a stable balance is restored. Since large marine carnivores like starfish, some arthropods, and particularly the cephalopods first appeared in strength in the Ordovician, this is probably another factor involved in the evolution of varied and numerous animals in Ordovician seas (Figure 4-4).

The different ways of life in the sea had been explored in general terms

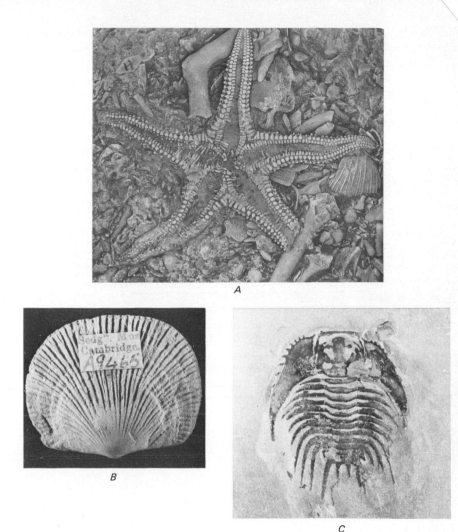

Figure 4-4 Paleozoic fossils. (a) *Promopaleaster*, an Ordovician starfish, (photograph courtesy of Jon W. Branstrator); (b) *Plaesiomys*, an Ordovician brachiopod; (c) *Leonaspis*, a Silurian trilobite. (Photograph courtesy of E. N. K. Clarkson.)

by the end of the Ordovician. There were floating photosynthesizing microorganisms, the plant plankton, on which food chains depended; they were eaten by microscopic floating animals, the animal plankton, probably consisting mainly of larvae of larger animals. Suspension feeders lived on plankton; carnivores and scavengers roamed the sea floor and the water column; bacteria decomposed organic remains and provided food for sponges. On the sea floor, algae were eaten by snails and sea urchins. In fact, the development of algae-eating animals largely kept algal mats out of normal sea water after the early Paleozoic. So their fossil remains (stromatolites) became very rare, and they are now found only in supersalty conditions where most animals cannot live. After the Ordovician, many areas of warm, shallow, clear water were occupied by reef dwellers such as algae, bryozoans, and corals, although Ordovician reefs were not as large, as varied, and as populous as modern reefs.

Although we have explained in general terms why each of these ways of life might have developed, we have not yet suggested why the *timing* should have occurred the way it did, to produce in succession the Ediacaran and Cambrian faunas dominated by deposit feeders, then the Ordovician faunas dominated by filtering suspension feeders. To solve this problem, we must first understand a revolution of thought that has transformed the science of geology since the early 1960s.

SUMMARY

R. B. Clark has made up a theoretical model for the early evolution of complex animals. Most of them evolved from wormlike animals that learned to burrow *into* Precambrian sea-floor sediments and in so doing evolved an internal body cavity or coelom. Precambrian soft-bodied animals, mainly wormlike, are preserved in rocks about 600 to 700 million years old.

About 600 million years ago, many animal groups evolved skeletons. This time marks the beginning of the Paleozoic era, and is an important biological and ecological boundary in the history of life on Earth.

The early Paleozoic era has Cambrian faunas which were mainly mud-eating animals. In the Ordovician period, which followed next, many marine animals came to filter food from the water. Large carnivores also evolved in the sea, so that marine biology became very complicated by the end of the Ordovician period.

FIVE

CONTINENTAL DRIFT AND THE HISTORY OF LIFE

CONTINENTAL DRIFT

The idea of continental movement is not new. The first full-blooded attempts to prove it took place in the early years of this century. But over the past ten years the evidence from different branches of geology has become so overwhelming that almost all geologists are now convinced that we live on a dynamic earth whose geography and climate can change slowly over millions of years as continents move over its surface.

The outermost layer of the Earth is the *lithosphere,* which forms the continents and ocean floors. It is extremely thin in comparison with the diameter of the Earth (about 2 percent at most), but it behaves like a cover of eight or so rigid plates rather than a pliable plastic scum (Figure 5-1). Internal flow in the Earth's mantle, probably convection driven by the Earth's internal heat, shifts the rigid plates of the lithosphere in relation to one another.

Only three possible kinds of movement can take place. *Two plates may drift apart.* If this happens, a rift or crack forms between them and molten material wells up the crack to form igneous rock. Because of the heat released near the rift, an expanded ridge bulges up. If a rift happens to form first under a continent, a great area is uplifted in a broad plateau, like the great upland plains of Kenya and Tanzania along the East African Rift Valley. Igneous rock forms the floors of these continental rifts. However, as plates are drifted apart so that the rift comes to be in the middle of a newly formed ocean, the expanded ridge then rises from the ocean floor to make a great

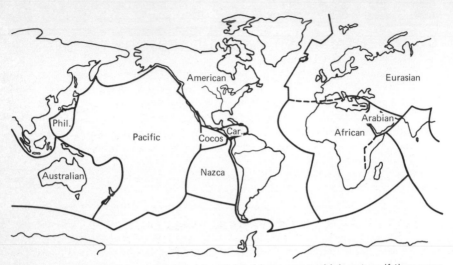

Figure 5-1 The Earth's surface can be divided into areas which act as if they were rigid plates.

submarine crest, really a long submarine mountain range, with a rift running down the middle of it (Figure 5-2).

Two plates may drift together. If the relatively thin and weak ocean floor is involved, it is pushed downward along the plate junction to form a deep trench in the ocean floor, and eventually it is remelted into the Earth's interior. But if two plates with strong and thick continental edges are drifted together, neither one can be pushed downward into the Earth; instead, their

Figure 5-2 New ocean is formed at midocean ridges; sometimes a continent is split apart when spreading begins.

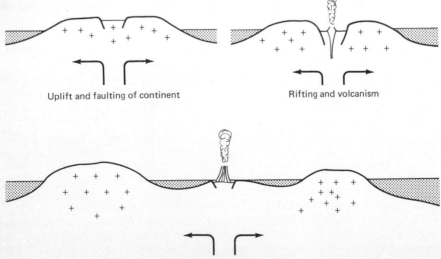

Uplift and faulting of continent

Rifting and volcanism

New ocean and midocean ridge

edges collide and crumple to form a large mountain range like the Himalayas. Finally, *two plates may slide sideways against one another,* forming a gigantic line of slippage that is called a major *transform fault* (like the San Andreas Fault in California).

Obviously, even if the movements of the plates are random (and we do not know whether they are or not), continents will be moved around on the Earth's surface, and oceans will form and be destroyed. As a result, the Earth's geography can be drastically changed over long periods of time. An average rate for plate movements is about 5 centimeters (2 inches) per year, so that thousands of kilometers of movement are possible over periods like 100 million years.

We can reconstruct the positions of continents and oceans in the past from several lines of evidence.

1. The similarity in shape of some opposing continental edges like South America and Africa shows that they were once in contact, but it does not give a timetable for their separation. By analyzing fossils on the two continents, fossils in the sediment on the new sea floor that formed between the separating continents, or by radiometric age dating of the rocks forming the ocean floor, we can construct a timetable for continental separation.

2. New lava flows often freeze so that magnetic minerals in the lava align themselves with the Earth's magnetic field. Any later movement of continent or ocean floor will move the magnetic minerals off this alignment. By reconstructing the position of the magnetic pole to which magnetic minerals "frozen" in a rock are pointing, it is possible to see how far they have moved in terms of *angles.* By comparing many data points, it is possible to calculate actual *distances* through which the rocks have been moved.

3. Many ancient mountain belts are lines along which plates once collided. Radiometric age dating is necessary to determine the time span over which a mountain belt was formed.

Naturally, in every reconstruction of global events, all these different lines of evidence are examined. In particular, it is vital to realize that unless the Earth has been expanding, the total amount of surface is always constant. Thus if new ocean floor is forming at a rift, old ocean floor carrying important evidence about the past is being destroyed somewhere else. We have to be increasingly wary about reconstructing events in the early history of the Earth, because so much of the ocean floor recording those events has been destroyed. In fact the oldest rocks found on the present ocean floors are only about 150 million years old.

CONTINENTS AND LIFE

The positions of the continents are vitally important to the life on and around them, because geography largely controls climate, and climatic factors are very important to living things. Thus near the poles of the Earth an organism must face six months of daylight and six months of darkness; food may be very abundant in the short summer, but almost absent in the long dark winter. Very few organisms can tolerate these strong seasonal changes, and so we say that life in these high latitudes has *low diversity*—there are not many *kinds* of organisms, although there may be large populations of those that do live there, like the fur seals of Alaska, or the penguins of Antarctica, or the black flies of the Canadian Arctic.

Near the Equator conditions are much more stable, and temperature, sunlight, rainfall, and food supply may not vary much from one part of the year to another. There is little or no seasonal change. In this sort of environment many different organisms can coexist, each specialized for a very particular way of life. The amount of food available to each animal may not be great, but it is dependable in this stable environment. Fresh vegetation grows year-round for leaf-eating monkeys in a tropical jungle, for example. Life in such low latitudes tends to be very diverse; hundreds of species of trees and insects can live in a square mile of jungle, and hundreds of species of molluscs can live along a few miles of tropical shore.

But this simple picture can be vastly altered by the distribution of continents over the surface of the Earth. For example, on the modern world there is a very great seasonal change during the year in the climate of India and Southeast Asia because they are part of a large continent. Winds blow out of Asia regularly during the winter when pressures are high in the interior of the continent, so that the winters are colder than one would expect from the latitude. In summer, when the sun's heat in the continental interior generates low pressures, winds are drawn in off the ocean to produce torrential rains in a great monsoon. There are dramatic seasonal changes in temperature, in rainfall or salinity, and particularly in food resources for the organisms on land and in the shallow seas along the shores, even in very low latitudes. Relatively few species flourish in these seasonal "continental" conditions, and they must develop ways of life to help them survive these changing environments. By contrast, in "oceanic" areas like the Caribbean islands, temperatures and food resources stay much more constant throughout the year, especially since they are near the Equator, so that life is much more stable. Even in high latitudes, "oceanic" areas like Japan or Britain have fairly mild climates. To make this contrast even more important, the climates of strongly seasonal continental areas are usually variable from year to year as well as from season to season. For example, the monsoons of India sometimes fail completely, causing widespread starvation for humans and animals.

In the past, when the movement of crustal plates acted to join Earth's continents into one or two "supercontinents," the climate of the world would have been strongly "continental," with strong seasonal variations in temperature and food resources, even near the Equator. On the other hand, when continental masses were broken up so that the world had many smaller continents, the world climate would have been much milder and "oceanic," stable in environmental factors except near the poles, where seasonal changes are always powerful. In a stable "oceanic" world with mild climates we would expect to find a great number of specialized animals relying on stable food resources.

Another factor comes into play. When the world is greatly split into separate continents, each continent or large island will tend to have its own sets of plants and animals, on shore and around its coasts, like the marsupials of Australia or the lemurs of Madagascar. In an "oceanic" world, where there are in any case many specialized animals, the existence of many separate geographic provinces of animals increases the variety of the world's life several times.

From time to time, it seems, there have been periods when some large continent has broken up into fragments, each fragment separated by a newly forming ocean with a midocean ridge along its center. When plate movements are particularly active, the midocean ridges may add up to quite a considerable volume of material. The ocean water displaced by these ridges would then be pushed up and away to flood the edges of the continents. Since there is a very large area of low-lying land along most coasts, a rather small rise in sea level can have drastic effects, causing shallow seas to cover vast areas which had been continental plains. If this happened, the world's climate would become much milder (because seawater has a moderating effect on climate) and there would be great new areas for marine animals to colonize. Even on land, although there would be less living space for plants and animals, the climate would be fairly gentle and stable, and a large diversity of forms could coexist.

In 1970 James W. Valentine and Eldridge M. Moores of the University of California, Davis, suggested that one overall result of the movements of crustal plates over the Earth's surface would be to control or to specify the diversity of life on Earth at any time. The implication of their suggestion is that major events in the history of life did not happen by chance, but were strongly influenced and perhaps even controlled by the great events that molded the Earth's crust by drifting continents around and creating and destroying oceans between these continents.

It follows that no one interested in the history of life can afford to ignore tectonics, the branch of geology that deals with movements of the Earth's crust. As a first example, we can study events toward the end of the Precambrian era. In the late Precambrian the continents were joined into a large supercontinent. At about the beginning of the Cambrian period this broke up into at least four separate continents, and by Middle Ordovician times there were probably several smaller masses spread widely over the Earth's surface. There was thus a great change in world geography, climate, and food resources during this time.

When food supplies are unpredictable, it is poor policy for an organism to rely on floating plankton for food, because the supply will be strongly seasonal, and periods of plenty will be followed by periods of starvation. The only comparatively reliable source of food in the sea in these conditions is in sediment on the sea floor, which is always present as a rich organic layer. Thus in an unstable world it is good policy to be a deposit feeder.

These ideas fit together to explain the changes in the fossil record in late Precambrian and early Paleozoic times. The large supercontinent of the late Precambrian must have had a very seasonal climate, and in fact we found that the Ediacaran fauna was composed very largely of deposit feeders, and that it was spread over most of the world, instead of being split into separate provinces, each with its own separate set of animals.

The Ediacaran animals were mainly wormlike, but in the Cambrian the development of skeletons was associated with the evolution of animal groups feeding on the *surface* of the sediment. Most Cambrian animals were deposit feeders still, and Cambrian faunas were dominated by mud-digging arthropods, the trilobites. But some animals adopted suspension feeding near the sediment surface (sponges and brachiopods, for instance). This change in habitat and feeding method may indicate that conditions were becoming more stable to some extent, along with the breakup of the Precambrian supercontinent. Suspended food in the water may have become a little more reliable because seasons would have been a little milder. Thus the evolution of hard skeletons in the Cambrian may have been an indirect result of continental movements which affected world climate. Furthermore, Cambrian fossils can be divided into three separate *geographical* groups, which suggests that separate ocean basins had formed.

When conditions are stable, there is a dependable food resource floating in the water, plankton, and in these conditions the best policy is to capture this food before competitors do. Thus, as the Ordovician period began, with the world now having widely separated continents, an "oceanic" geography, and presumably stable food resources, there was a great explosive evolution of suspension feeders, often modified to feed high in the water. Thus bryozoans built strong colonies, echinoderms developed long stems, and graptolites became free of roots on the sea floor and took on a floating way of life. The presence of so many animals exposed and reaching out into the water in turn provided a stable food supply for newly evolved wandering predators and scavengers like starfishes and cephalopods (Figure 4-4).

Thus in the late Precambrian and early Paleozoic, the major events of the biological world are associated with the major events in world geography in a way that makes sense for the first time. The ideas of Valentine and Moores were only put forward in 1970, and already they have solved problems which had been worrying paleontologists for a hundred years or more. We shall see more examples of this new insight as we progress through this account of the history of life.

SUMMARY

Great plates of the Earth's crust move slowly in relation to one another. Continents and oceans can change their sizes and positions. Changing world geography can set off great climatic and biological events over many millions of years, and we might expect to see such changes in the fossil record.

When there are many separate small continents on the Earth's surface, it is an "oceanic" world and life should be very diverse; when there is one supercontinent and one superocean, the world's climate is called "continental" on land and in the shallow waters where most marine animals live. Not many species live in this kind of world.

Altogether, the history of continental movements and world geography should control part of the history of life. This is so in the early Paleozoic. Separating continents should have encouraged life to become more diverse —and it did.

PALEOZOIC LIFE IN THE SEA

FOSSILS AND STRATIGRAPHY

Fossils can be studied in two different ways. One may simply study the different *kinds* of fossils (like collecting postage stamps), or one may study the different ecological *roles* they played in the biological world of their time. From both these viewpoints, the major features of evolution of life in the sea were completed by the end of the Ordovician, 450 million years ago. All the phyla of organisms we would expect to find in the marine fossil record were present, from sponges to chordates. Ecologically, all the major ways of life in the sea had developed, from mud flat to reef. Evolution did not stop, of course, and there were many changes during the rest of the Paleozoic and afterwards; but most of these later changes in marine animals were replacements rather than innovations, and were in any case variations on a theme rather than dramatic shifts of body plan or of ecological role. For example, the early stalked echinoderms of the Ordovician became extinct quite quickly, but were replaced by others which survived until the end of the Paleozoic. The early crawling molluscs of the Cambrian were replaced by more complicated spiral-shelled snails and by bivalves.

It is interesting to compare an Ordovician set of shore-living animals with a Devonian set from the same environment. Although the animals inhabiting each are very different, their ecological roles are very close.

As an example of variation on a theme, the Paleozoic cephalopods are fascinating. They all had chambered shells with a built-in buoyancy and

balance system like a submarine. They were probably all wandering opportunistic animals, crawling and swimming about on the sea floor, eating other
animals dead or alive, as scavengers and carnivores. Yet their sizes ranged
from a few millimeters (as young or small individuals) to as much as 30 feet
long. Their shells were long and straight, or tightly coiled into a spiral;
smooth or ribbed; with disklike or globular shape. Presumably the "adaptive
zone" of cephalopods (that is, the range of different roles they could play in
the life of the sea) was wide enough to include a great variety of shapes and
sizes, each efficient within its own particular way of life. At different times, a
particular shape would be better suited to the environment than another, and
a wave of extinct forms and newly evolved forms would be deposited in the
fossil record for us to puzzle over. Similar fluctuations in detail occurred in all
fossil groups, so that there was continuous evolution and a continuous
change in fossil faunas. This is the basis for *stratigraphical paleontology,* a
branch of geology which is concerned with establishing the *sequence* of
events in the deposition of the fossil record.

Stratigraphical paleontology is based on two principles. One is *superposition:* a new deposit of sediment is always laid down on top of pre-existing
surfaces of sediment or rock, so that in any pile of sedimentary layers in their
original positions, the older layers are on the bottom and the younger layers
are on the top. The other is the observation that *evolution is irreversible,* that
there never has been *precisely* the same set of organisms inhabiting the
world at more than one time. Thus any given time is marked, in principle, by a
set of deposits which contains a *unique* set of fossil remains of the organisms
living in the world at the time. If we can study these fossils carefully enough,
get them in the right order by using the principle of superposition, and set up
a list of particularly important ones, then we can work out a foolproof
sequence of "guide fossils," "index fossils," or *zone fossils,* whose occur-
rence will be absolutely sure evidence of the relative age of any given
sedimentary rock.

There is no way that stratigraphical paleontology can give the age of a
rock in absolute numbers of years, because we simply work out a sequence
of fossil collections. But at the present time we can distinguish more quickly,
more exactly, and more economically between neighboring parts of the fossil
record by using fossils than we can by using radiometric age dating. In
practice, of course, the age of a rock is determined in as many ways as
possible. By combining paleontological evidence with radiometric age dates,
we now have a good idea about the absolute ages of groups of fossils, and
this gives us clues about rates of evolution.

Unfortunately, stratigraphical paleontology is much more difficult in
practice than in theory, and its strengths and weaknesses are both related to
the fact that it is based on living things, which are always contrary beasts to
deal with. Particular organisms live in particular environments, so that one
would not normally find a living sand crab in a muddy bottom, or a fossil sand
crab in a mudstone. Organisms living in a particular environment form *facies
fossils,* where a *facies* is a set of rocks laid down under specific environmental conditions (for example, reef limestones in a "reef facies"). Facies
fossils are very good for reconstructing the biology and ecology of ancient

environments. But for fossils to be useful for stratigraphical paleontology, they should be found in all kinds of sedimentary rocks. The best "guide fossils" are those that lived for a geologically short time, and are numerous, widespread, and easily recognized. Floating or swimming organisms make the best guide fossils. They are more widely spread than other kinds, for they drop into all kinds of sediment after death. In the sea, floating animals like graptolites and foraminifera make good "guide fossils," whereas on land, plant pollen and spores blow for long distances and are equally valuable, especially as they may also blow out to sea and be preserved as fossils in marine facies.

Stratigraphical paleontology allows us to set up a sequence of fossils that in turn defines a relative time order for rock layers. Over the past hundred years a set of units has been developed to arrange these sequences in a convenient form. *Eras* like the Paleozoic era last for hundreds of millions of years; *periods* like the Cambrian may last for tens of millions. Periods are subdivided into *stages* of five to twenty million years, and *zones* are only one or a few million years long. There are three great eras divided into twelve periods from the Cambrian onward (see Table 6-1).

Early geologists intended that periods and eras should represent im-

Table 6-1 Geologic time scale

Era	Period	(Epochs)	Time began[†]	Symbol	
CENOZOIC	NEOGENE	(Pleistocene) (Pliocene) (Miocene)	2 25	TTT	
	PALEOGENE	(Oligocene) (Eocene) (Paleocene)	 65		
MESOZOIC	CRETACEOUS		136	K	
	JURASSIC		190	J	
	TRIASSIC		225	Ŧ	
PALEOZOIC	PERMIAN		280	P	
	CARBONIFEROUS	(PENNSYLVANIAN)	325	ℙ	C
		(MISSISSIPPIAN)	345	M	
	DEVONIAN		395	D	
	SILURIAN		430	S	
	ORDOVICIAN		500	O	
	CAMBRIAN		570	€	
PRECAMBRIAN			4600	p€	

*In North America the Mississippian and Pennsylvanian are regarded as periods, and the Paleocene, Eocene, etc. are regarded as epochs. In the rest of the world, the Mississippian and Pennsylvanian are usually combined into the Carboniferous period, and the Cenozoic "epochs" are regarded as "periods". These differences don't affect the realities of geology, but they worry people who are legal perfectionists.

[†]Figures in this column represent millions of years.

portant changes in the world's biology, and they hoped that important changes took place at the same time all over the world, making it easy to draw "timelines" in rocks on a worldwide scale.

If the early geologists were correct, we should expect the boundaries of periods (say, between the Cambrian and Ordovician) to reflect directly or indirectly some important event in continental movements that took place about the same time. In fact, this generally works out: for example, the end of the Cambrian period marks the explosive evolution of suspension-feeding organisms which came to dominate Ordovician seas. The continents were separating at this time.

Eras mark even greater crises for the world's organisms. Thus the Paleozoic began as skeletons were evolved for the first time, and ended with a dramatic extinction of many marine animals. The Mesozoic ended with the extinction of the dinosaurs and the ammonites, which populated land and sea respectively for more than 100 million years.

New ideas about the effects of continental drift on life finally give us the hope that we will be able to *explain* what each major period means in terms of its unique fossil remains, instead of simply making a catalog of fossils to be learned by paleontologists. In fact, explaining the events marking the boundaries between periods is one of the exciting challenges facing the geologists of the 1970s.

STABILITY AND CRISIS IN THE PALEOZOIC

The Paleozoic was a long era, which means that there were nearly 400 million years without a major biological crisis. Once the marine ecology of the world became established in Ordovician times, it remained fairly stable until the Permian. Some groups of animals were replaced by others which may have performed the same ecological function in a different manner. This general continuity did not extend to land environments, which were still being exploited and developed for the first time toward the end of Paleozoic.

There were some important biological events that affected marine ecosystems during the Paleozoic, and they seem to be related to continental movements as sea-floor spreading rafted crustal plates over the Earth's surface.

The breakup of an old Precambrian supercontinent resulted in a very "oceanic" world in the Ordovician, as we have seen, and a very varied fauna evolved to take advantage of numerous ecological opportunities. Later continental movements could only act to decrease the "oceanic" character of world geography, and some of the early Paleozoic continents collided and became locked together into larger units. Thus a "proto-Atlantic" ocean closed in Silurian times and North America locked together with northern Europe. This collision formed a great mountain chain whose eroded roots now form the mountains of Scandinavia. The British Isles also collided with North America, forming mountains whose roots are now hills in Scotland and New England. Altogether, by the middle Paleozoic the continents had been rearranged into only three major clusters (Figure 6-1).

These were *Gondwanaland* (consisting of the southern continents South

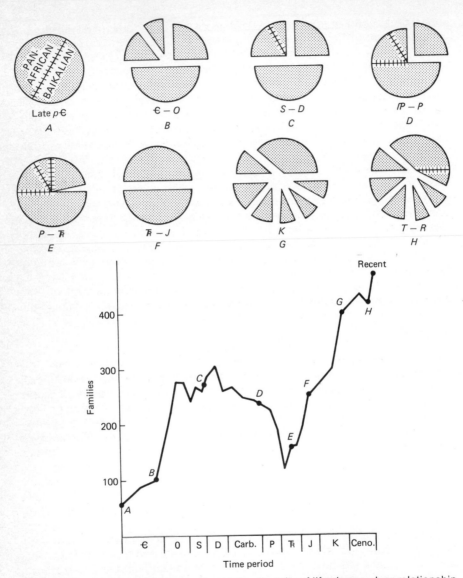

Figure 6-1 Continental movements and the diversity of life show a clear relationship from Cambrian to Recent times. For time symbols see Table 6-1. (After Valentine and Moores, *Journal of Geology* 80, 170, 180. Copyright © 1972 by the University of Chicago.)

America, Africa, India, Australia, and Antarctica); *Euramerica* (North America and Europe); and *Siberia* (Siberia and possibly the rest of Asia). These major clusters then also collided and merged late in the Paleozoic. Siberia and Euramerica collided along the line of the Ural Mountains of Russia in the Permian period, and together they are known as *Laurasia* (a northern supercontinent). Laurasia and Gondwanaland collided to form mountains in central Europe in the Pennsylvanian period.

The world, then, became generally more "continental" after the Devonian period as the number of separate continents decreased, and the number of separate ocean basins between them became fewer and narrower. Inevitably, the world climate would have become more extreme, seasonal, and "continental," and the supply of food resources from plankton would also have become much more seasonal and unreliable. In a reversal of the Cambro-Ordovician sequence of events, the world must have become much more severe for organisms relying directly on suspended food material like plankton, and instead the favored groups of organisms would again become those relying directly on detritus, the deposit feeders. In turn, the disruption of food chains relying directly on plankton must have severely upset the predators that fed on suspension feeders, so that carnivores would not have been favored in the deteriorating late Paleozoic periods.

THE PERMIAN EXTINCTION

By the end of the Permian period all the continents had collided and locked together to form a single land mass known as *Pangaea*. With only a single continent, the world's climate would then have been as extreme and seasonal as one could imagine.

The resulting Permian extinction was probably the most dramatic biological decline in all Earth history. Within five million years or so, whole groups of organisms were wiped out. Those particularly hard hit were the brachiopods, bryozoans, and crinoids (suspension feeders), and the corals and cephalopods (carnivores). Other groups also suffered extinctions (the last few trilobites finally became extinct, for example). Suspension feeders and carnivores rely on a steady supply of food and so must have been particularly vulnerable. Their devastating Permian extinction can be seen as a natural response. Land faunas were not so greatly affected, as we shall see in a later chapter.

Various catastrophic theories have been proposed to explain the Permian extinction, ranging from sulfur poisoning of the sea to suffocation from lack of oxygen. None of these theories accounts for the differential extinction of some ecological groups, which is easily accounted for by the idea that continental movements control the style of world ecosystems. But there is a serious problem still to be faced. The continents collided and merged into one over a period of 100 million years or more, from the Devonian to the Permian; yet the devastating extinctions occurred mainly during the last 5 million years of the Permian. Why was the extinction so sudden?

A geographical reconstruction of the Permian world (Figure 6-2) shows that the supercontinent Pangaea was not compact, but had a gigantic oceanic indentation on its eastern side, in tropical latitudes. This oceanic gulf has been named *Tethys*. With the continents all joined, the world ocean would have been 18,000 miles wide round the Equator, and the result would have been very unusual conditions within the Tethyan tropical gulf. A permanent westerly equatorial current would have poured warm water

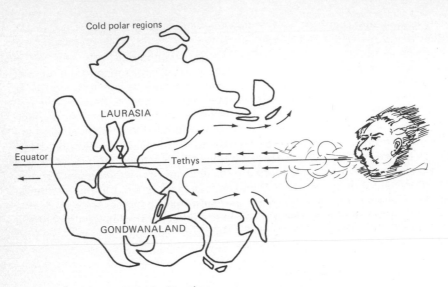

Figure 6-2 Geography and ocean currents in the Permian world.

continuously into it, so that here at least, in offshore environments like islands and fringing and barrier reefs, conditions would have been quite uniform from season to season. In the fossils of the Permian Tethyan ocean, we find some of the richest reef faunas that have ever existed on Earth, with hundreds of species inhabiting a single reef. Elsewhere in the world's seas, Permian faunas are rather sparse in numbers of species, indicating much more seasonal conditions.

At the end of the Permian, the special conditions giving uniform seasons in the Tethyan gulf must have been somehow interrupted. The Tethyan reef faunas were wiped out very rapidly as a result, thus completing the extinction that had already largely taken place in the rest of the world. In the Canadian Arctic, for example, Michael J. Fisher finds a very great change in spores and pollen at the end of the Pennsylvanian period. The suddenness of the Permian extinction is thus the result of special local conditions in the Tethyan gulf, which allowed late Paleozoic reef faunas to flourish there after the rest of their world had been lost to them.

What sudden event overtook the Tethyan reef faunas? Probably it was triggered by a retreat of the sea. When continents are actively moving, midocean ridges push water out of the ocean basins to flood the edges of the continents and create broad, shallow seas (see Chapter 5). When all the continents joined together into a giant supercontinent at the end of the Permian, the seas drained back off the continental edges. In the Tethyan gulf where there were very rich reefs, a retreat of the sea would have been the final blow that finished off the last of the rich Paleozoic faunas.

Evidence for the nature and timing of the Permian extinction has been gathered from a series of famous and exotic fossil outcrops, ranging from Texas to the Mediterranean, and through Turkey, India, Pakistan, China, and the Far East. Some of the best outcrops are apparently in south China, and they should provide a treasure house of information in years to come.

SUMMARY

Evolution is irreversible, so the fossil record should never repeat itself. In theory, geologists should be able to set up a time scale in which "guide fossils" would always be in the same relative position. By using this principle, the Earth's rock record has been divided into convenient units for easy reference. The breaks where major boundaries have been drawn occur where fossils changed rapidly. Such biological changes might reflect changes in world climate or geography resulting directly or indirectly from plate movements.

The Paleozoic era was long and marine ecology did not change a great deal during it. At the end of the Permian period all the continents had merged into one, and there was a great extinction of marine life as a result.

SEVEN

EARLY VERTEBRATES

VERTEBRATE ORIGINS

We are vertebrates ourselves. We are most familiar with vertebrate biology, and we identify readily with other members of our own species, and with pets and domesticated and wild animals that are vertebrates. So it is usually easier for us to understand the biology of extinct vertebrates than other organisms. Yet vertebrates are rather rare as fossils, though the skeletons of fossil whales and dinosaurs dominate museum halls. Living vertebrates are far outnumbered by arthropods on land and molluscs in the sea. Vertebrate skeletons often fall apart after death, because their bones and scales are held together only by muscle and cartilage which easily rot. Bone itself is easily dissolved and crumbled. Furthermore, many ancient vertebrates lived on land where their bones were easily scattered or destroyed, not covered by sediment and preserved, as is more likely to happen in the sea. In vertebrates more than any other group, there are unfilled "gaps" in the fossil record, although we can usually reconstruct a reasonable story to make up for the lack of direct evidence.

The most serious failure of the fossil record is the lack of direct evidence about the first of the vertebrates. Vertebrates must have evolved from invertebrates, which occur much earlier in the fossil record. There is no direct fossil link between the two, so it seems that the immediate ancestor of vertebrates was probably soft-bodied.

Vertebrates belong to the phylum Chordata, which includes some animals without hard backbones. All chordates have a long, straight internal support, a *notochord,* which acts as a stiffening structure; in most chordates it is surrounded by the vertebrae of the backbone. The notochord is in fact a very specialized organ constructed from muscle fibers. When it contracts, it becomes very stiff, but it can be relaxed to give some flexibility. Probably in very early or very primitive chordates this alternating stiffness and flexibility aided in swimming; it provided a rod for muscles to pull against, while allowing the animal to flex in a swimming motion.

Normally, chordates have a long axis along which the notochord runs, with some pairs of sensory organs concentrated at the front, thus defining a head. In most chordates the main nerve center is housed in a special cavity; that is, the brain is encased in a skull. Vertebrates probably evolved from a chordate ancestor without a backbone built around the notochord. There are two living chordates which might be similar to the vertebrate ancestor.

Tunicates ("sea squirts") seem to be very unlikely chordates when they are adult. They are essentially small boxes with two openings through which water passes to be filtered clear of microorganisms inside the animal. But the larval tunicate is very different: it looks like a tadpole and is clearly divided into "head" and "tail," with a notochord passing along the main axis. The tunicate larva moves by a swimming motion of the tail.

Amphioxus is a lancelet, a small animal which lives close inshore in sand, filtering suspended particles from seawater. Water is taken in at the front, filtered, and passed out through gill slits. A notochord runs the whole length of the body, and muscles act against its stiffness to produce bending and swimming motions of the body.

We do not know whether the ancestral vertebrate resembled either *Amphioxus* or a tunicate larva. But it is a reasonable guess that the first fishes evolved from a small, soft-bodied, active chordate, with a notochord and gills, probably making a living by filtering particles from seawater. It is unlikely that this kind of chordate could have been very successful before Ordovician times, when world conditions first favored the rapid evolution of suspension-feeding ways of life.

Earliest Fishes The earliest vertebrate fossils are some small bony phosphate plates from Early Ordovician rocks near Leningrad in Russia. These are certainly heavy fish scales, and they are found in sands which also contain glauconite, a mineral only formed in seawater. Some scientists argue that vertebrates first evolved in fresh water, but there is really no good evidence for this. In the United States, the earliest fish scales are found in the Middle Ordovician Harding Sandstone of Colorado, which is also thought to have formed in a shallow sea.

Two points are interesting about these earliest fish scales. First, no internal bones have been discovered. Apparently, then, the first fishes were living inside plated boxes, with only a notochord for internal stiffening. The scales must have been protective, perhaps against powerful invertebrate predators like the first nautiloid cephalopods, or some large arthropods. Second, the scales were made of phosphate, which may have been a way to set up a phosphate store as well as a protective plating. Phosphate is a necessary mineral for life, but is often in short supply in the sea. Thus the first fishes solved two problems as they evolved external scales. They also created a new problem for themselves: inside their heavy phosphate boxes, they must have been relatively immobile, rather slow heavy swimmers near the bottom.

The first fishes had no jaws which could open and close the mouth by a hinging mechanism. Instead, they fed through a small slit or opening at or under the tip of the snout. Their food must have been small and easily digested without any chewing. Living jawless fishes, the hagfish and the lamprey, have a very different biology; the lamprey, for instance, is a parasite on other fishes. So in order to understand the biology and evolution of early jawless fishes we have to rely largely on their fossil remains, not on their living relatives.

There was only one group of jawless fishes from the Early Ordovician to the Late Silurian. They were usually small, and had a large flattened head shield made of several bony plates, with the eyes at the sides. It is fairly obvious that they were bottom-living, scooping food off the sea floor. Some of them had plates round the mouth that could have been extended out into a shoveling scoop. Because of their rigid plated bodies, they must have propelled themselves only with the tail fin, which again indicates that they were not sophisticated swimmers.

The jawless fishes quickly developed an amazing variety of shapes in the Late Silurian and Early Devonian, and must have taken on new ways of life (Figure 7-1). This was a time of rapid evolution of marine life, invertebrates as well as vertebrates. Among the jawless fishes, there were body shapes adapted for swimming in open waters, for sucking sediment through a tube, and for scooping tiny floating plankton from surface waters. The head shield became specialized for gliding through water in some cases, as a delta-wing aircraft glides through air. In other cases there was a large sensory organ developed on the side of the head, possibly suggesting a complex pressure or electric detector for use in murky waters. Some jawless fishes are very much like modern catfish which live in streams, and there is strong evidence from the sedimentary rock record that jawless fishes quite early became adapted to fresh water. The problems associated with the change from salt to fresh water were mainly biochemical and did not leave traces on the hard parts, so that we cannot get much evidence directly from the fossils.

Early in the Devonian, the jawless fishes had achieved a wide variety of ways of life in spite of their great disadvantage—they were confined to eating very fine particles, either floating plankton or sediment on the sea floor. By a paradox which reappears several times in the fossil record, it was at their peak that they were faced with overwhelming competition from a new source—the fishes that had evolved fully functional hinging jaws. This spelled the end of the jawless fishes, except for a few survivors which have rather strange ways of making a living in modern seas.

THE EVOLUTION OF JAWS
Fishes have bony struts (gill arches) supporting their gills, and it is likely that at some point the flow of water across the gills was increased by evolving a pumping action, flexing the forward pair of gill arches. Gradually this pair of

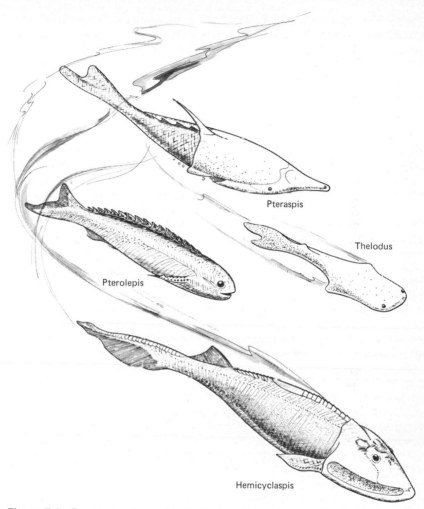

Figure 7-1 Reconstructions of early jawless fishes, to show the variety that evolved among this group. (From E. C. Colbert, *Evolution of the Vertebrates.* Copyright John Wiley & Sons, Inc.)

gill arches was modified into a pair of hinged jaws, so that the mouth could open much wider, increasing water flow over the gills, accommodating larger mouthfuls of food, and providing a jointed pair of biting, chewing, and crushing organs (jaws) fitted with piercing and cutting implements (teeth).

Jawed fishes first appeared in the Late Silurian, and therefore must have evolved from the earlier jawless fishes. But when we look at the few living jawless fishes, a problem arises: they all have gills *inside* the gill arches (essentially in the mouth), whereas all living jawed fishes have gills *outside* the gill arches. It is very difficult to imagine any evolutionary intermediate between the two, although many attempts are made in current textbooks. The answer to the dilemma is probably that the wide range of *early* jawless fishes included forms with gills inside and gills outside the gill arches. But only those with gills *outside* the gill arches had the evolutionary option of transforming gill arches into jaws without interfering with the function of the gills. Thus an apparently minor difference in mouth structure was eventually of tremendous evolutionary importance.

The first jawed fishes which appeared in the Late Silurian were surprisingly modern in their appearance. They were the *acanthodians,* and were small, light, spiny, scaly fishes with large eyes, paired fins, and a streamlined shape. The jaws had many small teeth. Unlike jawless fishes, the acanthodians were well equipped for fairly rapid and well-coordinated darting and swimming in open water. Their fins were largely supported by strong spines, which may mean that they were not used for propulsion but rather for balance and steering; swimming power came largely from body and tail. Although acanthodians are sometimes called "spiny sharks," they are not related to sharks at all, and since they were only a few inches long they could never have been large predators like sharks. They were numerous in the Late Silurian and Early Devonian in both fresh and salt water, but afterwards it seems that they were outcompeted by the later and more advanced groups of jawed fishes, and they slowly became extinct.

A second group of jawed fishes also appeared in the Late Silurian, a short time after the acanthodians. The *placoderms* were usually heavily armored, with a heavy shield covering the head and body. This was in two parts, with a pair of hinged joints between the head shield and the body shield. There were two groups of placoderms, the arthrodires and the antiarchs. The arthrodires were sometimes huge, perhaps more than 30 feet long, and they dominated some Late Devonian fish faunas. A large lower jaw had sharp bony plates which sheared against similar plates on the upper jaw, making a savage cutting device. The arthrodires could gape very widely by lifting the skull and upper jaw on the joint with the body shield at the same time that they dropped the lower jaw. The strong armor did not extend backward very far, so that a powerful swimming action could have been operated by a long muscular body and tail.

Early arthrodires were small, a little flattened, and with rather weak jaws, but during the Devonian they developed into fast-swimming killers up to gigantic size (Figure 7-2). In this process, the body shield became relatively shorter, and more of the body must then have been available for propulsion by flexing. At the same time the jaw mechanics improved, giving a wider gape, and possibly pumping water more efficiently through the gills.

The antiarchs followed a different way of life. They were more heavily armored, far back along the body, and were flattened rather than streamlined like the arthrodires. The eyes were usually small and placed on top of the flat head shield, indicating a life on the sea floor. A small mouth with a weak jaw completes the picture of a slow-moving fish looking rather like the early jawless fishes. But antiarchs are interesting because they had powerful jointed fins looking superficially like arms, presumably used for "walking" or "poling" the fish along the bottom. They were thin and pointed and would have made poor paddles. The overall impression of antiarch biology is that they must have been mud diggers like the early jawless fishes.

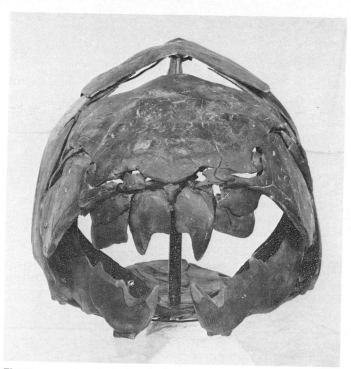

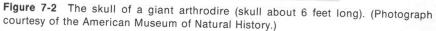

Figure 7-2 The skull of a giant arthrodire (skull about 6 feet long). (Photograph courtesy of the American Museum of Natural History.)

A particularly interesting feature of antiarchs is evidence that they had lungs, indicating air breathing. We shall return to this question, but it is enough to say at this point that many early fishes seem to have been provided with lungs and capable of breathing air.

The placoderms as a whole had heavy armored plating. Some forms show traces of an internal skeleton made of cartilage rather than bone. Placoderms are not closely related to any important living group of fishes, and so they were not the ancestors of bony fishes or of sharks. They became extinct very suddenly at the end of the Devonian period, even though they were the most savage and powerful fishes of the time.

SHARKS AND BONY FISHES

Living fishes fall into two categories, cartilaginous and bony fishes. Very little needs to be said about the former group, which includes sharks and rays. The internal skeleton is made of cartilage rather than bone, and there is no internal swim bladder. Fertilization is internal, and the female lays eggs in special leathery cases. Sharks and rays are notoriously successful animals, and there is no basis for the opinion that they are somehow "primitive." Modern sharks have complicated sensory systems, including an excellent sense of smell and an electric sensing system; they have complex behavior patterns. The only reason they might be called primitive is because they appear in the fossil record in the Devonian, and some modern forms are essentially the same as the Devonian ones. Rays are a more recent offshoot from sharks. Modern sharks are all marine, but sedimentary rock evidence suggests there were some freshwater sharks in the late Paleozoic.

The major problem in tracing the evolution of sharks is that the cartilaginous skeleton is not often preserved. Most fossil remains of sharks are teeth and spines, which do not help much in assessing the size, shape, or ecology of the entire animal.

The bony fishes are divided into three groups:

Actinopterygii   (ray-fin fishes, dominate modern seas)
Dipnoi   (lungfishes)
Crossopterygii   (lobe-fin fishes, extinct except for coelacanths)

All three groups appeared in the Early Devonian, probably evolving from different acanthodians, although there is no good evidence about this. Bony fishes generally do not have heavy surface armor, they have a bony internal skeleton, and they all had lungs when they first evolved. In most modern bony fishes, the lungs have been modified into a swim bladder which helps to maintain buoyancy in the water, but in the living lungfishes the lung is still used for breathing air.

Actinopterygians Actinopterygians or ray-finned fishes have scales that differ from other fishes, so that they can be recognized even from fragmentary fossil remains. They have fins which consist of dominantly radial elements or rays. Their first fossils are known from Early Devonian freshwater deposits, but they evolved rapidly and were seagoing before the end of the Devonian. Since then they have dominated marine and freshwater environments of the world. Naturally, these fishes have evolved all kinds of ways of life, but they are typically lightly built fast swimmers in open water.

There is a complicated evolutionary story in their later history, but in general the evolution of the actinopterygians is a sequence of events which led to a lighter skeleton, lighter scales, and radical improvements to skull and jaws to make them lighter and more efficient. In this group, air breathing was probably discontinued early in their history as they invaded the sea, and the lungs were modified into an air bladder for buoyancy control in swimming up and down in the water.

Of all other marine animals, fishes were most directly in competition with cephalopods, which were already numerous in Devonian seas when the actinopterygians invaded from fresh water. Even today, modern squids are as efficient as fishes in acceleration and maneuverability, hydrodynamic and metabolic efficiency, sensory discrimination, intelligence, and behavioral complexity. But in general it seems that the fishes have succeeded in driving most cephalopods off the continental shelves of the world and into deeper oceanic waters.

Lungfishes The Dipnoi, or lungfishes, and the Crossopterygii, or lobe-fin fishes, are quite closely related. They had the same kind of scales, small eyes, and long snouts (relying on smell as well as sight to a greater degree than actinopterygians), and in particular they had similar fins. The fin had a set of longitudinal bones as well as radial elements, so that there was a long jointed central axis controlling fin movements. This probably gave the fin a slower and more powerful propulsive beat, in contrast to the fast responsive flutter of an actinopterygian fin (which can be seen in the nearest goldfish bowl). The first lungfish can only be distinguished from early crossopterygians by its skull structure, but this is important and shows that there had been evolutionary divergence before the Early Devonian when they both appear in the fossil record.

Lungfishes have apparently always been freshwater fishes, and have always had a body structure which suggests fairly slow swimming. Teeth were gradually reduced during evolution to large flattened plates for crushing small molluscs and crustaceans, and scavenging plant and animal debris. Lungfishes apparently evolved to live in seasonal rivers and streams. They faced special problems associated with drying out—the lack of oxygen in the water, and the loss of water altogether. As a result, modern lungfishes can in some cases burrow into a muddy stream bed and nearly "switch off" their body processes during a dry season, to emerge during the next rainy season. Their lungs help them to survive periods of low oxygen level in the water. In many ways, these specializations have allowed lungfishes to survive in conditions which other fishes could not tolerate, but there are only a few marginal environments to which lungfishes are well adapted. There are only three living species, which inhabit rivers in Africa, Australia, and South America. Lungfishes are "living fossils" in the sense that fossil "lungfish burrows" have been found in Carboniferous rocks, suggesting that their present survival strategy has carried them precariously through more than 250 million years.

Crossopterygians The crossopterygians are an amazing contrast to the Dipnoi in spite of the similarities we have noted. As the lungfishes gradually adapted toward a sluggish way of life in seasonal streams, the crossopterygians evolved toward a very active mode of life as inhabitants of large rivers and swamps. In fact, they were the dominant predators in Devonian freshwater environments, being long and powerful streamlined fishes. Only a few ever invaded the sea, but one of these groups, the *coelacanths,* has a surviving form which lives now in deep water off South Africa. The most important group of crossopterygians, however, was the *rhipidistians,* adapted for fast swimming in Devonian freshwater environments. They had a special set of joints in the head, by which the snout could be raised and lowered on a joint with the skull, while the lower jaw could open independently. This system served two useful purposes. First, raising and lowering the snout alone changed the mouth volume, allowing water to be pumped through the gills, or air to be pumped in and out of the lungs, without moving the lower jaw. Second, in a predatory chase at high speed in shallow water, it might be dangerous to drop the lower jaw to swallow prey;

instead, the snout could be raised. Modern crocodiles have the same kind of problem, and the same kind of solution.

Rhipidistians were obviously strong swimmers, with long, flexible bodies. Their fins were set low on the body, and in active pursuit in very shallow water the fins could have helped the fish to plow through and over shallow mudbanks. The fins were probably most useful during rest periods, however, when they could have supported the weight of the body on a shallow bank to allow free movement of the rib cage in breathing air. Probably rhipidistians spent considerable time in this activity (or lack of it), basking, breathing, and digesting in sunshine and comfort, free from predators. Although we now know that rhipidistian fins developed into amphibian limbs, land-going would have been rather useless to a fast-swimming fish like a rhipidistian. The lobe-fin was useful to the fish only as part of its life-style as a shallow-water hunter. The rhipidistian rib cage was much stronger than that of a coelacanth or a lungfish, allowing continuous breathing while the weight was supported by lobe-fins in very shallow water.

In pursuing their shallow-water life, some rhipidistians gave rise to amphibians in the Late Devonian. This innovation eventually restricted those rhipidistians that remained in rivers and swamps, probably because the newly evolved amphibians were better predators in very shallow water than their ancestors, and essentially pushed them out of this habitat. The remaining rhipidistians in the Carboniferous and Permian tended to be larger and were adapted to larger rivers and lakes.

SUMMARY

The earliest vertebrates were scaly jawless fishes that lived in Early Ordovician seas. Presumably they evolved from a soft-bodied chordate ancestor which in turn evolved from an invertebrate.

Jawless fishes became widely varied by Early Devonian times, and lived in fresh- and saltwater environments. But one group evolved jaws in the Late Silurian, and jawed fishes quickly outcompeted most jawless forms.

Some groups of jawed fishes were very successful in Devonian times, but quickly became extinct. Sharks and rays, on the other hand, are still successful in modern seas.

Bony fishes are the most numerous group of fishes, and they dominate modern seas. Some Devonian freshwater fishes (the rhipidistians) eventually gave rise to land-going vertebrates.

EIGHT

FROM WATER TO AIR

INTRODUCTION

People usually write of "the invasion of the land" to describe how plants, invertebrates, and vertebrates left the sea. It is better to call this great step *adaptation to air*. Many marine organisms spend all their lives fixed to or crawling on a hard surface, just as many land organisms do. But life in the sea and on land differs greatly as a result of the different properties of water and air. Land organisms are bathed in gas (air) rather than liquid (water).

Air is less dense than water, and so *organisms weigh more* in air, causing problems in their physical support. We feel the same effect after getting out of a swimming pool. Seawater carries a great variety of dissolved or floating food substances, while air does not, so that organisms have *special problems in food supply* in an air environment. Life in air poses a *drying-out problem* because air is hardly ever saturated with water vapor. Drying out is a severe problem for adult animals, but is particularly *critical for reproduction* since reproductive cells are very sensitive to drying out. Oxygen and carbon dioxide behave very differently in air and water, so that an *adjustment in respiration* must be made. The refractive index of air is lower than water, demanding *changes in eyes*. Sound transmission is different in water and air, so that *hearing devices must change*.

Evolution by natural selection takes place by the gradual accumulation of beneficial changes of the DNA in the gene pool of a population. However drastic the changes required to transfer from life in water to life in air, they must have been accumulated very slowly, in a succession of populations which were all successful organisms (or they would have become extinct!). *None* of the problems listed above could have been solved overnight by the sudden appearance of a revolutionary mutation. The gene structure of an individual is in a very delicate balance, and large changes will almost always disrupt so many interlocking physical and biochemical characters that an unfit descendant will result.

For a sequence of populations to become adapted for life in air, all the problems would have to be overcome slowly and more or less at the same time, while the organisms were still well adapted for life in water. Only at the point in the sequence when all the problems had been solved adequately (if not perfectly) could the organisms have emerged into air for long periods. We must be able to reconstruct in our minds a sequence of adaptations which would have the end result of success in air, while at the same time explaining how the populations in question were still living successfully in water.

The evolution of animals is a very complex process. In an animal, different structures may act together to operate a particular function; thus lungs, blood circulation, heart, and rib muscles may all act together in the process of breathing. Naturally any change in one component may automatically involve change in all of them, and we have to take this into account when we examine sequences of fossils.

Evolutionary change seems sometimes to anticipate events. Thus fishes developed air breathing before they invaded the shore or the riverbank. Actually, this is inevitable, but we should clearly understand that they were not *deliberately* accumulating devices for existence in air or for crawling out onto land. They were adapting in response to the environment in which they were living, and *by chance* those adaptations permitted them to extend their range in a landward direction. *Preadaptation* is the term used to describe this kind of fortunate evolutionary sequence, but it does not imply any evolutionary forethought by the animals involved. Natural selection is entirely the product of a chance process (mutation) and is never directed. Possibly this is the most difficult part of the theory of evolution to accept, but it is a fundamental point.

We have to be very careful in putting together an explanation of the evolution of amphibians from fishes. It must fit the available data, it must provide a reasonable evolutionary pathway which makes sense in terms of the biology and environments of the later Devonian, and it must make both physiological and structural sense. In many ways, the fact that a good story can be pieced together is a triumph for paleobiological methods.

FROM FISHES TO AMPHIBIANS

Physiology   Animals which breathe take in oxygen (O_2) which is used to release energy for body processes by burning "food." Carbon dioxide (CO_2) is a waste product which is toxic because it dissolves easily in water to form carbonic acid (H_2CO_3). Animals can tolerate small concentrations of CO_2, but most of it must be passed out of the system. Both O_2 intake and CO_2 output are accomplished by passing body fluid (blood) very close to the body surface. If the concentration of the gas in the blood is very different from that outside, gas will diffuse through the body wall to equalize the concentrations. Usually, gas exchange sites are localized, richly supplied with blood vessels, and have very thin walls, as in lungs and gills.

In most parts of the sea it is easy to get rid of CO_2, which dissolves and diffuses away very readily in water. Usually seawater is well oxygenated, so that it is easy to obtain fresh oxygen by passing deoxygenated blood, full of CO_2, through the gills: O_2 and CO_2 diffuse in opposite directions across the gill surface. Most living fishes use this system. But in some water bodies, particularly in lakes, streams, and partly enclosed lagoons containing rotting vegetation, the oxygen level in the water may be rather low, temporarily or permanently. Low oxygen levels are extremely dangerous for fishes breathing only through gills, since it is difficult to set up a large enough intake of oxygen through the gills.

Some modern fishes which face habitats like this "breathe" air to provide them with oxygen, because normal air is reliably 21 percent oxygen, twenty times as much as in fully oxygenated water. Modern fishes that seem to "breathe" air actually bite off an air bubble and absorb it through the back of the mouth. In this way they take in both O_2 and a small amount of CO_2. They cannot *exchange* O_2 and CO_2 in the mouth, because the buildup of carbonic acid in an air bubble in the mouth would be too great. Instead, they get rid of the CO_2 *at the gills,* because it diffuses away easily in the water.

Thus air-breathing fishes do not operate the same exchange system as we do, although their system is efficient in its own way. Its major disadvantages are: first, a free swimming fish can only visit the surface to breathe every now and again, so that breathing rates cannot be very rapid; and second, the air bubble causes some buoyancy which must be compensated by change in the swimming style of the fish.

Air breathing by most Early Silurian and Devonian fishes probably arose as they invaded low-oxygen freshwater environments, and led to the development of a special pouch behind the mouth cavity (the lung) to accommodate air. A large lung would allow a greater time between breaths. Most of the fishes which returned to the sea no longer needed the lung (seawater is usually rich in oxygen) but they modified it into a swim bladder to aid in buoyancy control. But those fishes which remained in Devonian rivers and swamps kept and improved on the lung. In their method of oxygen intake, these Devonian fishes were already fully adapted to air.

The system for *losing* CO_2 must be different in air, because gill filaments stick together out of water and cannot operate efficiently. Some fishes get rid of CO_2 by diffusing it away through the skin, and in small living amphibians like frogs this so-called "skin respiration" is the main method of losing CO_2. But this process has severe drawbacks. A large area of body surface must be used for gas exchange, so that the skin must be thin and soft, with no protective armor. As a result, most living frogs and toads rely on stealth and agility, camouflage, or poison as defenses. A thin skin adapted to lose CO_2 also loses large amounts of water vapor, so that a drying-out problem can develop; living amphibians are usually found in moist environments. Finally, since volume increases more quickly than surface area in a growing organism, the skin surface becomes inadequate for CO_2 loss from a large animal; living amphibians are small.

Another way of losing CO_2 is used by a few fishes, and by reptiles, birds, and mammals—it is *to breathe rapidly.* A mouthful or a lungful of air is partially exchanged and then breathed out before carbonic acid builds up to dangerous levels in the lungs. The oxygen of the air inhaled is never completely exchanged, and a great volume of air has to be moved in comparison with the amount of oxygen absorbed. But respiration is performed in one localized region of the body, and large body size and high metabolic rates become possible. The skin is not devoted to CO_2 loss, and can develop a wide variety of modifications, such as scales, armor, hair, and feathers. Solving the problem of CO_2 loss in air was a major accomplishment, and it could have been acquired slowly by the gradual evolution of a rapid-breathing system.

Rapid breathing usually requires the mechanical pumping of air in and out of the lungs. A great variety of pumps has evolved among vertebrates. In most mammals and reptiles, muscular movements of the rib cage and/or diaphragm serve to move air. But in crocodiles, which spend much of their time resting with all their weight on the rib cage, lung volume is changed by jerking the liver rapidly to and fro. In small amphibians, a complex set of air movements involving motions of the floor of the mouth can be seen inflating and deflating the throat pouch of a frog or toad.

In paleobiology, we must be very careful in making comparisons between living animals and their fossil relatives, because these are sometimes very misleading. Thus it is futile to compare a lizard weighing a few grams with a dinosaur weighing 40,000 kilograms, although the comparison is often made. In the same way, the small living amphibians may be quite misleading if we compare them uncritically to the first amphibians which evolved 350 million years before them.

The first amphibians were rather strong-boned, and had *scaly* bodies of reasonable size. It is practically certain, therefore, that they did not use skin respiration, but instead they probably used the rib cage to alter lung volume for rapid breathing. It is not clear when the first frogs, toads, and salamanders evolved, but there is general agreement that they evolved later than the fairly large scaly early amphibians we have just described.

Structure Rhipidistian fishes were long, fast-swimming, and streamlined, with powerful lobe-fins set rather low on the body. The first amphibians were long-bodied animals with a long powerful tail carrying a fin, and they had four limbs set rather far toward the sides of the body. The bones of the amphibian limb, essentially in a pattern one-two-several-many like our own limbs, can be matched with the bones of the rhipidistian lobe-fins. Although we do not have the full sequence of fossils which would precisely show the evolutionary pathway, there must have been a gradual change from fin to limb, and we have to explain *why* this change would have been an advantage for the first amphibians.

Early amphibians spent much of their time in water. The body and tail would have been powerful swimming devices, aided a little by the limbs. In the water they would have been rather like small crocodiles, with sharp pointed teeth, eyes set rather high on the head, and a well-streamlined shape. For all these characters, the rhipidistian fishes are reasonable ancestors. The movement power of rhipidistians came mainly from body and tail, with the lobe-fins probably acting only as aids to maneuvering. If the rhipidistians pursued prey into very shallow water, they could presumably plow effectively through mud and shallow water by body movements, but their breathing would be jeopardized because a lot of weight would fall on the rib cage, preventing it from inflating and deflating fully. In these circumstances, the lobe-fins, acting as supporting props, could have raised the body enough to allow full breathing.

Gradually, in those rhipidistians which habitually hunted in very shallow water, the lobe-fins might have evolved to make them stronger, supporting the body more directly through stronger bones at shoulder and hip. The fins developed into firm ridged implements for grip on a muddy or sandy bottom, rather than smooth pliable surfaces for swimming control. Thus lobe-fins would develop into limbs and feet, propulsion devices with toes for pushing against a firm bottom, and not simply props for support in a vertical direction (Figure 8-1).

This way of life would eventually result in the gradual evolution of rhipidistians into amphibians. Girdles of bone developed at shoulder and hip, acting as a secure foundation for the pushing limbs, and tying them to the rib cage and backbone for maximum leverage. Thus the animal adapted toward walking on land while still being fully capable of swimming in water by means of its body and tail (Figure 8-2).

This hypothetical sequence of events is supported by other indirect evidence. Rhipidistians have a skull and jaws adapted to open wide in a short head by raising the snout and lowering the lower jaw. As amphibians evolved, the snout lengthened and the upper jaw became fully attached to the skull. This is possibly connected with the pursuit of smaller prey, in which a smaller gape but a longer reach would be important. At the same time, early amphibians developed greater flexibility of the joint at the back of the skull, forming a real "neck" for the first time, and giving the potential for rapid sideways snapping of the snout. The smaller prey of the first amphibians

Figure 8-1 Rhipidistian fishes and early amphibians were very similar in the way their body motions were used to give movement. Evolution from fins to feet was not some kind of miracle.

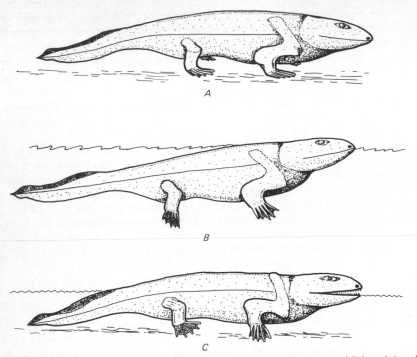

Figure 8-2 Reconstruction of *Ichthyostega*, the first known amphibian; (*a*) walking, (*b*) swimming, and (*c*) plowing over a shallow mud bank.

could have been small fishes in very shallow water, or early flightless insects on the water surface or on vegetation.

Fishes detect sound waves in water through their *lateral line*, a row of pressure-sensitive receptors running down the side of the body. The ear in fishes is primarily a fluid-filled balancing organ, inside the braincase, but presumably it can "hear" vibrations passing through the skeleton. In early amphibians, the lateral line system was used under water, but a new sensory device was evolved for hearing in air. A jawbone passing close to the ear in fishes evolved to lie in one of the unused gill slits, and in amphibians came to transmit sound waves received on a membrane from the outside inwards to the inner ear through a gap in the skull. The old gill slit continued into the throat as the Eustachian tube which we humans still possess.

Eyes must have undergone drastic changes to adapt to vision in air, but in the fossil record we can see traces of only one change. The evidence of tear ducts, necessary in air to moisten and clean the eye surface, is preserved in grooves on the bone next to the eye of early amphibians.

Reproduction Amphibians are tied to water for reproduction. Desert-dwelling forms like the spadefoot toad must wait in frustration for a rare rainstorm to provide puddles of water. Amphibian eggs are fishlike, produced in large numbers by the female, and usually fertilized externally by the male shedding sperm on them. The eggs have little or no protection against drying.

Modern amphibian larvae hatch early into the water, and must feed and survive as tadpoles in water before going through a drastic metamorphosis which produces a miniature adult adapted for a truly amphibian life. There is no direct way of knowing how metamorphosis evolved in living amphibians or when it happened, but it may not have been so drastic in early amphibians which were still very similar to rhipidistian fishes in form and structure. Possibly the modern amphibian metamorphosis was perfected only with the evolution of frogs and toads, much later than the Devonian. Larval amphibians from Pennsylvanian and Permian rocks are miniature adults except that they have gills for CO_2 loss in water.

In reproductive biology, the amphibians have not really succeeded in adapting fully to air. Probably this has been the greatest single factor in the relative lack of success of amphibians, although we should remember that some early amphibians evolved extremely quickly to solve the problem, and were so successful that we now class them as a separate group, the *reptiles*.

THE FIRST KNOWN AMPHIBIANS

Sometime before the Late Devonian, rhipidistian fishes gradually evolved into amphibians. The first known amphibians are a group of fossils from eastern Greenland, of which the best known is *Ichthyostega*. There is no doubt that they were descended from rhipidistians, because of similar features like the skull bones, the vertebrae, the palate and teeth, and the bone sequence in the limbs.

Yet *Ichthyostega* is a tantalizing fossil. It was quite large, about three feet long, with a heavy skeleton. The shoulder and hip girdles and the rib cage were very strong, so that *Ichthyostega* was well adapted for walking on land. There was a long tail with a fin, and the body was well streamlined, so that it was also capable of efficient swimming. *Ichthyostega* was propelled on land by limbs and feet, in the water by the tail, and in very shallow water by both, much like a modern crocodile (Figure 8-2).

In these features it is a good transitional form between fishes and later land vertebrates. But other characters make it clear that *Ichthyostega* was not the direct "missing link" between fishes and all other vertebrates. For example, the backbone was rather weak in structure compared with other land-going vertebrates—the strength of the body was maintained by the strong barrel-chested rib cage, not the vertebrae. This makes good engineering sense, but it is not the method used by other land-going vertebrates.

THE ENVIRONMENT OF EARLY AMPHIBIANS

World geography in Late Devonian and Early Carboniferous (Mississippian) times included a large continent in the North Atlantic region, Euramerica, now broken into Europe, Greenland, and North America. Its southern shore lay along the equator, with rivers and lakes draining southward to the ocean. There were no dry-land plants at this time (at least there is no fossil record of them), so that large deposits of red sands and muds were washed into lakes and shoreline sediments. This has caused the continent to be called "The Old Red Sandstone Continent," and it has been interpreted frequently but wrongly as a giant desert. Although it was probably barren of inland vegetation, its climate could have been fairly mild, hot, and wet. Fishes were abundant in its rivers and lakes, as shown by the fossil record.

Along the southern shore of Euramerica and inland, equatorial latitudes imply tropical conditions with little seasonal change, like Malaysia or the Amazon Basin today. This is important, because temperature extremes are usually more severe in air than in water. Conditions along the continental Devonian shore and inland must have been ideal for animals and plants to evolve from water to air through a very humid swampy environment (Figure 8-3).

The first land plants are known from Late Silurian rocks. Plants faced problems similar to those of animals in adapting to air. The tremendous surface area of plant leaves, spread out to gather light for photosynthesis, posed a desiccation problem solved by the development of a watertight waxy outer coat. Nutrients are available on land only from the soil, so plants required root systems and internal transport systems for nutrients and for water. Support also demanded the firm attachment of some kind of root, and large size depended on the development of hard supporting tissue like wood. The reproductive system of plants also had to change to avoid drying out the reproductive cells.

As in animals, most of these problems were solved only very gradually. The first land plants were very simple forms, without any complex supporting structures and without extensive root systems. They probably lived only in wet areas at the edges of swamps. Later plants developed many of the food-supply and support systems needed for larger size, but the reproductive system was still dependent on dampness to avoid drying out.

Plants received great advantages by invading the air medium. Fixed plants in the sea are restricted largely to the narrow zone along the shoreline where light penetrates the sediment-laden, wave-churned water. Invasion of the land immediately opened up a vast area for colonization. But the greatest advantage of air is that the ultraviolet radiation from the sun is much more intense in the absence of a water screen, and this allows greater photosynthetic production of food.

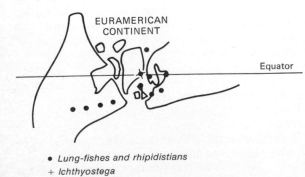

EURAMERICAN CONTINENT

Equator

• Lung-fishes and rhipidistians
+ Ichthyostega

Figure 8-3 Devonian lungfishes and rhipidistians and *Ichthyostega* lived along the southern edge of the Euramerican continent in tropical latitudes. (Data from Carroll, 1969.)

In Late Devonian times, the "land" flora probably lived only in swampy areas, but within that habitat there were plants varying from mosses up to large, strong tree ferns. The invasion of swamps and mudbanks by plants must have contributed heavily to the amount of rotting debris in ponds and bayous. In turn this led to lower oxygen levels in the water (oxygen is used up in decay processes), and encouraged the development of air breathing among contemporary fishes and arthropods.

As plants gradually extended their range into swamps and onto banks and levees, arthropods of various kinds apparently followed this food source. The first "land" arthropods are known from a swamp deposit from the Middle Devonian. Two separate groups had already solved the problem of survival in air—the ancestors of insects and the ancestors of spiders and scorpions. Middle Devonian insects were wingless, but by the later Carboniferous huge dragonflies had evolved, and the full use of the air medium had begun.

Plants like tree ferns and horsetails dominated Carboniferous floras, and had developed strong woody tissues and good root systems which allowed them to grow tens of feet into the air. Competition for light encouraged the growth of thick forest stands. The accumulation of vast quantities of plant debris in swamps and deltas led to the formation of great coal beds, now exploited by man from the Ural Mountains to the Appalachians. The Carboniferous forests were growing in damp soil and water, where oxygen levels were very low. Like modern trees growing in the same kind of environment (such as the Okefenokee Swamp in Georgia), Carboniferous plants had roots which did not extend vertically into the ground, but spread out sideways. Trees built this way are easily felled by wind and storm, and there must have been a great quantity of rotting wood and plant debris lying on the surface, forming an ideal habitat for microorganisms and small arthropods.

There are many different habitats within a swamp or delta. Strong-flowing streams, tidal creeks, or major branches of a river would have been ideal habitats for fast-swimming rhipidistian fishes. More restricted lagoons and bayous would have provided habitats for smaller fishes and aquatic amphibians of various sizes. Mazes of fallen and rotting tree trunks and fern thickets would have provided damp shady places for arthropods of many kinds, and for both large and small amphibians that were adapting toward more terrestrial conditions. These amphibians could have been very varied in their biology—large, stealthy predators and smaller, quicker predators. Probably there was a kind of land-based food chain on the banks and levees, where large amphibians ate smaller amphibians which ate insects and grubs which in turn were living off microorganisms or plant material.

A great many amphibians of varied size, structure, and habits populated this complex world, beginning in the early Carboniferous. It is unlikely that *Ichthyostega* was the only type of Late Devonian amphibian. We may eventually expect to find other smaller amphibians in Late Devonian rocks, with the body stiffened in the "usual" amphibian way, by vertebrae rather than rib cage, and with the kinds of varied ways of life that might have given rise to all the different Carboniferous amphibians which we now know and think we understand.

AMPHIBIAN EVOLUTION

It is very difficult to work out the details of early amphibian evolution. There are five different groups, and in each group there was rapid evolution in different directions. Basically, the amphibians evolved in two major directions. One was toward a more terrestrial life. Their bodies became shorter (twenty to twenty-five vertebrae in front of the pelvis), and their limbs became longer and stronger. Backbones were very strong, with the vertebrae so arranged that they resisted twisting and flexing. The "land" amphibians included *Eryops*, which was several feet long and probably lived like a crocodile, basking and resting above water, but feeding on fishes and animals at the water's edge. *Seymouria* was a carnivore about the size of a large dog; it is interesting because it had some features which are reptilian. *Seymouria* is not ancestral to reptiles, but it shows that several groups of amphibians were evolving reptilian characters independently. At this time, even the most "terrestrial" amphibians were still limited to swampy environments because of their reproductive limitations. A truly dry-land ecology did not evolve before the Permian period at the earliest.

The other main direction of amphibian evolution was toward a life in water. Amphibians could not have competed with rhipidistian fishes in swift predatory life in rivers and deep lakes, but some of them probably invaded lagoons and bayous. In these amphibians the body became longer (thirty to forty vertebrae in front of the pelvis), the backbone was adapted for flexibility rather than strength, and the limbs became weaker and shorter. Aquatic amphibians were probably rather sluggish swimmers and floaters, but there are very few clues about their biology. They survived well into the Triassic period, longer than the early terrestrial amphibians.

There are some early amphibians which are not well understood. Some were very snakelike, with about 200 vertebrae and no trace of limbs. They might have been aquatic or truly amphibian, but they must have moved in a snakelike way. Others were partly snakelike, and partly plain puzzling, since they had very flattened heads with enormous lateral horns of unknown function.

SUMMARY

Before organisms could live in air, they had to solve many problems. Air does not support weight, it is lacking in food, it is dry, and it carries light and sound in ways different from water. Its oxygen content is richer than water.

Organisms had to be able to survive conditions in air before they could leave the water. Probably the great advantage for animals on land was the rich supply of oxygen in the air, compared with low oxygen levels in swamp waters. Rhipidistian fishes were well designed for hunting small fishes and insects in very shallow water. In following this way of life, they seem to have put together the set of characters which allowed them to invade land environments. At this point in evolution we recognize the appearance of amphibians. They seem to have lived in warm, humid swamps along the edge of a Late Devonian continent.

NINE

REPTILES

THE ORIGIN OF REPTILES

Amphibians successfully solved many of the problems associated with living in air, but their reproductive system never became fully adapted to air. Amphibian eggs are comparatively small and their outer membrane is not resistant to drying out, so amphibians must breed and lay their eggs in water or in very damp places. Living *reptiles* have successfully overcome this problem, and reptiles were the first truly dry-land vertebrates. The critical adaptation was the development of the reptilian egg, although there were several other features that formed part of this evolutionary breakthrough.

The reptilian egg is enclosed in a tough membrane which in turn is encased in a leathery or calcareous shell. These two layers allow some gas exchange of oxygen and carbon dioxide with the air, but they effectively resist drying. They also give the egg some strength, for it must retain its shape lying on a hard surface in an air medium. Reptilian embryos develop inside these protective layers directly into miniature adults, unlike living amphibians. Inside the egg, as if in a spaceship, the embryo must have with it a complete life-support system. A yolk provides nourishment, and a special sac acts as a gas-exchange organ and as a waste-disposal unit. The embryo itself floats in a liquid-filled sac, the *amnion,* so that it develops and grows in a liquid pool which also acts as a shock absorber.

The reptilian egg is a very sophisticated system, and this means that it affects many other features of the biology of reptiles. The fact that the egg is encased in two membranes means that internal fertilization is compulsory—the egg must be fertilized before it is encased in shell. The egg must be much larger than that of a fish or amphibian, because it contains much more material. In turn, the reptile produces fewer eggs, so that each egg is more precious. Complicated devices for concealing and protecting clutches of eggs have evolved in many reptiles, as well as in birds. Because of the large size of the egg, very few reptiles are tiny. (Eggs must be laid somehow!)

We will never be able to reconstruct fully the evolution of the reptilian or amniote egg from the fossil record, since its development largely concerns changes in soft parts and in behavior. Like all evolutionary changes, it must have taken place gradually. A few living amphibians, for example, have developed a primitive kind of internal fertilization. Of course, those early reptiles which originally developed a rather crude waterproofing for their eggs have long since been eliminated in competition with more advanced reptiles. Most likely the transition from amphibian egg to reptilian egg was made in a very damp environment, where advanced amphibians could lay their eggs in very humid air. Here the protective membranes for a reptilian egg would be gradually evolved because they helped survival during dry periods and gave better support and physical protection for the embryo. As the developing embryo was not living freely in water filled with food and enemies, the special characters of the amphibian tadpole stage were not needed, so there is direct development of the reptilian embryo.

Since some amphibians evolved gradually by natural selection into reptiles, it is impossible to draw a definite line between the two except by choosing some artificial basis for a definition. We already know, for example, that the first amphibians were very unlike modern frogs, toads, and salamanders. In fact, in many ways early amphibians were more like reptiles than modern amphibians.

Reptiles are characterized by their reproductive system, but this is impossible to see directly in fossils. Instead, several skeletal features are used to define a reptile. They include minor changes in the position of the ear and changes in the jaw. It is only in later reptiles that really important differences from amphibian skeletal characters developed. Despite this difficulty, recent research has shed a great deal of light and common sense on the problem of reptile origins. We can now begin to understand it biologically.

The first known reptiles are from rocks of Early Pennsylvanian age, suggesting that they evolved from amphibians sometime during the Mississippian period. The earliest reptiles are called *romeriids* after Professor A.S. Romer of Harvard University, a famous vertebrate paleontologist. They are well known, thanks to a freak of preservation—their skeletons were found inside the hollow stumps of fossil trees still standing upright in Pennsylvanian rocks at various places in Nova Scotia, Canada.

The first reptiles were much smaller than their contemporary amphibians—they were about the same size as small living lizards, and like them in body proportions and posture (Figure 9-1). They probably ate insects and grubs, and they show enough differences from one another in details of teeth and limbs to suggest that they occupied a wide range of environments and/or

Figure 9-1 The early romeriid reptiles were very like living lizards in size and structure, so their biology may also have been much the same. (Reconstructions from Carroll and Baird, *Bulletin of the Museum of Comparative Zoology* 143, 1972, copyright by the President and Fellows of Harvard College.)

sought after a wide variety of prey. It certainly seems that an important fact about the origins of reptiles involved *small body size*. There are several good reasons for this, which are connected with movement, food, reproduction, and climate.

Amphibian adaptation toward a more terrestrial way of life would involve dependence on grubs, worms, and insects rather than on fish for food. An animal evolving in this direction would have to be small, first to maintain itself on this highly nutritious but small-sized prey, and second because capturing the prey might well involve a great deal of quickness and agility. A small body size would also be important for foraging after small prey concealed in awkward cracks and crevices, and for agility in a maze of plant and wood debris on the ground; a comparison with modern lizards makes sense here. A small, light body would also be most favored as limb structures evolved fully toward load bearing and quick movement on land.

The change in diet to insects, grubs, and worms was associated with a change in jaw mechanics. Rhipidistian fishes and amphibians operated a jaw designed for slamming shut on a prey. Early reptiles and a few small amphibians independently evolved a jaw designed more for holding, chewing, and crushing than for slamming. Coupled with their small size, early reptiles also had comparatively short skulls compared with most early amphibians. These features make sense in terms of capturing and holding small agile terrestrial prey. The early reptilian palate was designed to hold prey while the jaws shifted slightly to get a better grip.

As the amphibian egg began to adapt toward the reptilian egg, gas exchange through the membrane and newly evolving shell would have been easier if the egg were comparatively small (with a comparatively large surface area relative to volume). Only later, when the reptilian egg was fully perfected, would a large egg size have become possible. So the reptilian egg seems to have evolved first in an evolutionary line with small eggs (and small bodies).

Away from the damp swamp itself, temperature extremes must have been rather greater, both daily and seasonally. An ability to tolerate temperature extremes would have been important. Stable temperature is important to most animals because body enzymes work best at a certain temperature.

Mammals achieve this by precise thermoregulation at a temperature which is almost always above that of their environment. Reptiles do not have stable body temperatures like mammals do, but they are not "cold-blooded" either. They have a surprising degree of control over their internal temperatures by behaving in particular ways, such as basking in the morning and evening sun, doing "push-ups" to warm themselves, or hiding from the heat of the sun and the cold of the night or winter in crevices and burrows. For a group of very early reptiles adjusting to a land environment away from a swamp, temperature control would have been much easier at small body size—it would have been quicker to warm a small body in the sun, and easier to hide from heat or cold in cracks and crevices. When behavioral and physiological patterns and structures had been improved, then larger body size could be reached.

In summary, the knowledge we have of the amphibian-reptile transition shows that it was associated with evolution toward fully terrestrial life, and that it took place among small agile amphibians sometime late in the Mississippian period. In other words, the evolution to a fully terrestrial vertebrate was extremely rapid considering that the first known amphibian is Late Devonian in age. Some small reptilelike limb bones are known from Early Mississippian rocks, which may suggest that reptilelike amphibians evolved very quickly indeed.

A RABBLE OF REPTILES

The innovations involved in the evolution of reptiles from amphibians were so successful that the reptiles were able to expand into an immense variety of forms. They were exploiting a new range of habitats, and in doing so they took on a great variety of shapes, sizes, and biologies, very soon after the first small agile reptiles developed. The very success of these small lizardlike reptiles meant that they could provide a food resource for some larger, more powerful reptilian carnivores. In addition, some early reptiles evolved back into swamp and water, away from their land origins. Some others found new methods for temperature tolerance, and came to live still further from the swampy tropical or equatorial habitats of the early amphibians.

Most important of all, some reptiles became vegetarian, existing directly on plant material instead of exploiting it secondhand by eating vegetarian insects and grubs. This was perhaps the final step in the evolution of vertebrates into land environments, because it allowed them to reach large size and to live wherever plants were plentiful.

All these different adaptive pathways imply the development of a variety of reptilian types. In Permian times we can distinctly recognize no less than seven orders of reptiles, each with a different body plan and ecology. This radiation happened so quickly that we cannot yet see in detail how the various groups developed from the earliest reptiles.

There were three great lines of land reptiles. The *eosuchians* were the ancestors of lizards, snakes, and the strange New Zealand tuatara, *Sphenodon*. Eosuchians and their descendants have always remained small, and we do not have a good fossil record of their evolution until the Triassic.

Thecodonts were larger, more active predatory Triassic reptiles, and from them were descended dinosaurs, crocodiles, and birds. But the largest and most abundant animals of the Permian were the *pelycosaurs,* which later developed into mammallike reptiles and finally into mammals.

There are four groups of aquatic reptiles, but at present it is unknown how they evolved. The *mesosaurs* were an early group with modifications for a life in water eating small fishes. But the extinct *ichthyosaurs* and *plesiosaurs,* and the living *turtles,* have left few clues about their origins. Mesosaurs are only important because they seem to have been freshwater fish eaters, yet their fossils are known from both South Africa and South America. This fact has been used to help confirm that these two continents were in contact in late Paleozoic times as part of the supercontinent Gondwanaland.

Finally, there is another strange group of Permian reptiles, including the first land vertebrate vegetarian. It had a heavy body and flattened grinding teeth for dealing with tough plant material. Its evolution from the earliest reptiles cannot yet be traced from the fossil record.

PELYCOSAURS

Pelycosaurs were like romeriids, but they evolved to a large size, presumably because they had solved the various problems associated with diet, reproduction, and thermoregulation. Large Permian reptiles were either herbivores living on plants or they were carnivores living on other reptiles and amphibians. Large herbivores included *Edaphosaurus,* which was up to 11 feet long including the tail, with a broad barrel-shaped body, and probably weighed about as much as a small cow.

Another line of evolution among the pelycosaurs led to large, powerful carnivores. Here some teeth became large stabbing weapons, and at a maximum length of 11 feet, pelycosaurs like *Dimetrodon* were the dominant land animals of Early Permian times. The teeth of these carnivores showed a high degree of adaptation for different functions, in a primitive way rather like the incisors, canines, and molars of mammals (Figure 9-2).

Edaphosaurus and *Dimetrodon* both developed greatly extended vertebral spines which projected directly upward along the back. Covered in life by skin, these spines must have supported a large "sail." Probably the sail was a device for regulating temperature extremes in open country. Both animals were too big to be able to hide completely from heat or cold; they probably used the sail for basking in the morning and evening sun, and for radiating heat away during the midday period by turning the sail end-on to the sun. At night they would have been able to cut off the blood supply to the sail, and so retain heat in the body during the cold hours. In this way they might have been able to keep their inner body temperature within comparatively close limits, so that their enzyme systems could be "tuned" for maximum efficiency—in *Dimetrodon,* for active life as a predator, and in *Edaphosaurus,* for digestion of large quantities of vegetation. Running the body temperature at a constant value higher than the environment would have allowed faster body processes, particularly of digestion, which would have been advantageous to both these animals.

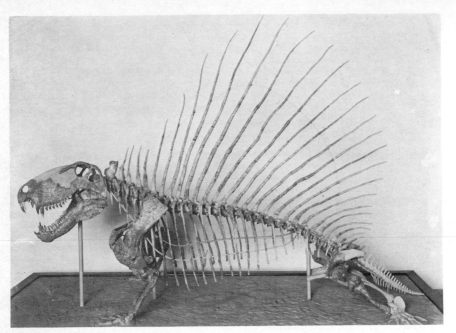

Figure 9-2 *Dimetrodon* dominated some Permian environments. It was carnivorous, and had a fantastic "sail" formed from vertebral spines, perhaps for heat regulation. (Photograph courtesy of the U. S. National Museum.)

MAMMALLIKE REPTILES—THERAPSIDS

By late Permian times the pelycosaurs had been largely replaced by their descendants, the therapsids. They evolved from the carnivorous pelycosaurs, but soon came to be herbivores as well as carnivores. All the therapsids had characters which allow them to be called "mammallike" in many respects. Like the carnivorous pelycosaurs, they had well-differentiated teeth, with distinct incisors, canines, and cheek teeth. A secondary palate separated the breathing passages from the chewing mouth cavity. This allowed a therapsid to breathe properly while chewing, something which most reptiles cannot do. This new adaptation was most important, because it had to be perfected before infant suckling could begin to evolve. Traces of follicles for whiskers have been found on the therapsid snout, implying that at least some hairlike growths had evolved on the skin. The lower jaw of therapsids was dominated by one bone, the dentary, although it had not yet reached the mammalian stage where the dentary bone alone makes up the lower jaw and the other bones have been modified into the mammalian middle ear.

The rest of the skeleton showed some changes. Therapsids tended to have stocky bodies and short tails, which may have helped them to retain heat. The limbs were still sprawling like those of lizards, though the skeleton was rather lightly built and therapsids would have been reasonably agile.

The variety of Permian land animals allows some tentative reconstructions of Permian biology. Organisms had spread far beyond the simple swamp ecosystems of the Mississippian and Pennsylvanian to take part in complicated land-based ecosystems. Everett C. Olson of the University of California, Los Angeles, has distinguished four different ecosystems in Early Permian beds of Texas:

1. In delta streams and lakes, the fauna was dominated by fishes and amphibians: sharks, rhipidistians, lungfishes and bony fishes, and aquatic amphibians.
2. In swamps there were lungfishes and more terrestrial amphibians.
3. In areas of lush lowland vegetation there were large vegetarian reptiles like *Edaphosaurus,* with carnivorous reptiles like *Dimetrodon* feeding on them. Away from the swamp, they both had "sails" to help regulate temperature.
4. Farther away from the swamp were small and large reptiles, using a food chain depending on arthropods to eat plants and provide food for larger animals.

In later Permian localities in the U.S.A. and the U.S.S.R., therapsids replaced pelycosaurs in land communities. Thus, by the end of the Permian, the dominant land animals were the therapsid mammallike reptiles: they occupied ecological positions with large and small body size, with eating styles ranging from insectivore to large carnivore and herbivore; and the stage seemed set for the emergence of true mammals as the dominant animals on Earth. There was indeed a revolution of world vertebrates in the Triassic, but it came from a completely unexpected direction.

SUMMARY
Reptiles were the first vertebrates to adapt fully to life on land. They evolved the amniote egg, in which a developing embryo could live as in a space capsule, protected from drying out but supplied with food, oxygen, and sanitary facilities.

The first reptiles seem to have been smaller than their amphibian ancestors. This is reasonable, because they would have been better adapted for hunting small insects on the forest floor of late Paleozoic riverbanks if they were small. The small early reptiles quickly evolved into an amazing variety of animals, which included the ancestors of swimming and flying reptiles, living reptiles, dinosaurs, birds, and mammals. By the end of the Permian period there were well-developed sets of land-going animals in ecological balance in environments from riverbanks to drier plains.

TEN

DINOSAUR EVOLUTION AND BIOLOGY

THE PERMO-TRIASSIC REVOLUTION ON LAND

Some idea of the dominance of mammallike reptiles on land can be gained by counting the variety of different known Late Permian vertebrates. The mammallike reptiles numbered 170 genera during this time, while all other reptiles put together came to 15. Even allowing for failures of the fossil record for smaller animals, or for too enthusiastic work on mammallike reptiles, this is an astounding disproportion.

Yet by the end of the Triassic, about 50 million years later, this situation had almost exactly reversed. Mammallike reptiles and mammals totaled seventeen genera, while there were then eighty three kinds of other reptiles. Pamela Robinson of the University of London has noted that this total reversal is not easy to explain, because it happened on a worldwide scale, independent of environment or geographical region, and the new reptilian groups filled the ecological niches of insectivore, carnivore, and herbivore that the mammallike forms had vacated.

The Triassic was a time when all the modern types of reptiles evolved. The three major groups of seagoing reptiles, the ichthyosaurs, plesiosaurs, and turtles, appeared; so did the direct ancestors of lizards and the tuatara, of crocodilians, and of dinosaurs. The first aerial vertebrate—a reptile—evolved, and mammals evolved from mammallike reptiles. This revolution was gradual rather than sudden, but it did result in a complete reorganization of the world's vertebrate faunas. Among the casualties were most of the early amphibians and reptiles. Two main arguments have been proposed to account for this revolution. One may be called climatic and physiological, the other structural. Probably both are correct, and together they acted to bring about the major faunal change.

The climatic and physiological argument proposes that there were rather different paleoclimatic situations in the Permian and Triassic. In both, the continents of the world were assembled together into a giant V, facing eastward, with Laurasia and Gondwanaland north and south of an equatorial oceanic gulf, called Tethys. Laurasia and Gondwanaland were joined at their western ends, so that Tethys was closed off on the west. In Chapter 6 we saw that the seas drained off the continents at the end of the Permian. Pamela Robinson argues that for most of the Permian, climatic conditions were comparatively mild, while Triassic conditions were more severe, with violent monsoon seasons bringing yearly wet and hot, dry conditions to the great land continents. She suggests that the lizardlike reptiles would be physiologically better adapted to survival in the hot, dry conditions of the Triassic, whereas the milder and damper Permian conditions were more favorable for the mammallike reptiles.

This may account for the success of *some* of the Triassic reptiles. For example, the rhynchocephalians were a successful group of Triassic reptiles, although only one species survives today, the New Zealand tuatara *Sphenodon.* Lizards of various kinds were quite abundant. But we have seen that the pelycosaurs and mammallike reptiles both had some modifications toward thermoregulation; these should have helped them to succeed, too, but they did not. So this climatic and physiological factor cannot be the whole explanation for the success of reptilian forms. In particular, it cannot explain the sudden appearance and success of the swimming *marine* reptiles—ichthyosaurs, plesiosaurs, and turtles.

Robert Bakker of Harvard University suggested that a *structural* innovation in one line of reptiles was instrumental in *their* success. Again, this is not the whole explanation, because it deals with only one group of reptiles. Bakker points out that the mammallike reptiles all had sprawling limbs; like living lizards, their upper limb bones projected sideways from the body in a rather awkward-looking arrangement. Among living animals at least, this type of limb is always associated with a fairly low body temperature, a fairly low metabolic rate, and a rather slow-moving way of life. Among living animals, whether they are mammals, birds or reptiles, the more erect the limbs are, the more active, agile, and energetic the animal is.

The Thecodonts   A new group of reptiles, the *thecodonts,* evolved in the Early Triassic. At first they were rather strongly built carnivores about the same size as the larger mammallike reptiles, with a similar clumsy sprawling walk. However, one of the thecodont groups evolved a lighter skeleton and became at least partly *bipedal.* Although they walked on four legs, they were capable of running fast on their hind legs alone, which were rather longer and stronger than the front legs. One of this group was *Euparkeria,* which was about 3 feet long, very lightly built, with a long, strong tail to balance it while running. It had a fairly long, lightly built skull with many sharp stabbing teeth, and was obviously a swift-running predator capable of catching and dispatching small animals.

Some later thecodonts showed great improvements over this early form. *Saltoposuchus* was a little longer and larger, but had hind legs much longer than the forelimbs and was probably bipedal most of the time. Most important, the limbs were set much more directly under the body, so that the feet fell directly in line while running. This must have had two major effects. First, the mechanical stresses of running fast must have been greatly reduced, and second, the trackway needed for running or walking was greatly narrowed. Possibly some small or young thecodonts hunted along narrow branches seeking insects, as modern chameleons do; they would have been greatly aided by this new limb structure.

In modern animals, more erect limbs indicate greater agility, speed, body temperature, and energy. Living tropical monitor lizards, for example, have semierect limbs and are active predators, with a preferred body temperature of 37°C, the same as many mammals. Triassic thecodonts are in fact very similar in size, proportions, and limb structure to monitors, and were probably as active or more active, since they had longer legs. Thus the thecodonts evolved what must have been a much more dynamic way of life than the mammallike reptiles, which were still slow and sprawling. The active way of life, with a preferred high body temperature, uses more energy, but this sometimes pays off. In the Triassic the thecodonts were faster to achieve the benefits of an active energetic life than the mammallike reptiles, and simply outcompeted them.

The replacement of mammallike reptiles by other reptiles was not catastrophic, but took place over millions of years. Some mammallike reptiles did survive to give rise to the true mammals, but they were certainly few in the Late Triassic, just as a few "primitive" mammals survive among many advanced ones on the present Earth. In general the active and dominant land vertebrates after the Triassic were the thecodonts and their descendants, the dinosaurs.

One can gather more evidence about the biology of thecodonts, because they have living relatives—the crocodiles and alligators. Truly modern crocodiles appeared at the beginning of the Jurassic period and evolved from thecodonts. The first crocodile was probably land-going, with strong, erect legs. Later crocodiles were much better adapted to water, and had a very important modification to the palate so that the nostrils at the very end of the snout became completely separate from the mouth cavity. They could then bite and chew under water.

The important point here is that we can look at living crocodiles in an attempt to understand the biology of thecodonts. Living crocodiles have a complicated circulatory system, with much better heart and lung modifications than other living reptiles. They have two ways of moving on land: a slow sprawling crawl, and a faster run in which the four legs are nearly vertical. One terrified scientist actually saw an 8-foot alligator running on two legs for 30 yards. It is quite possible that crocodilian features have been inherited from thecodont ancestors that were much more active than their living descendants (which are actually very lazy). Semierect limbs and high activity levels in thecodonts would have allowed them to outcompete the mammal-like reptiles in the Triassic, thus partly explaining the relative success of Triassic reptiles. Robert Bakker's ideas about this certainly seem very persuasive, though we should remember that they cannot explain the success of *all* reptilian groups in the Triassic.

THE PROBLEMS OF DINOSAUR BIOLOGY

There has never been anything like a dinosaur. The largest living land mammals (elephants at 6 tons) are only about one-fifth of the weight of the large dinosaurs, and the closest living relative of the dinosaurs, the crocodiles, only weigh a ton or so. Some dinosaurs approached 100 feet in length (the longest living snake is about 30 feet long). All this means that a paleontologist has no living animals for comparison in trying to understand dinosaur biology.

At present, most reconstructions of dinosaurs are based on comparisons with living reptiles like lizards, so that dinosaurs are usually reconstructed with sprawling limbs, and are considered to be "cold-blooded," i.e. unable to control their body temperature. Their braincase was small in comparison with body size, so that the general impression in popular books is that dinosaurs were slow, clumsy, sluggish, and stupid. Mammals, on the other hand, are associated in one's mind with activity and intelligence. In fact, it was even suggested at one time that dinosaurs became extinct because the early mammals stole all their eggs!

Yet there must be something wrong with this impression. Dinosaurs really did rule the world, and they did so for more than 100 million years. During all this time there was no mammal bigger than a cat; even lizards were larger. In fact, no adult dinosaur was *smaller* than 3 feet long and 20 pounds in weight. But as soon as dinosaurs became extinct, mammals evolved to great size, variety, and abundance. These are *facts,* and they do not fit the idea that dinosaurs were slow and stupid; it is clear that dinosaurs were so successful biologically that mammals were confined to a way of life with small body size for 120 million years.

THE EVOLUTION OF DINOSAURS

The Triassic thecodonts evolved into four major groups of reptiles. *Crocodiles* have been discussed; they were heavier thecodonts that eventually adapted to life largely in water. The evolution of the great *gliding reptiles* will be discussed later.

Thecodonts were also the ancestors of the two great groups of dinosaurs, *saurischians* and *ornithischians.* Both were very successful and each took up many different ways of life. The major difference between them was in the structure of the pelvis, and this is related to different ways of raising the hind legs as the animal walked or ran. Both ornithischian and saurischian dinosaurs had bipedal and quadrupedal members.

The two dinosaur groups evolved in the Late Triassic. Both groups were then still similar to thecodonts; they were fast-moving animals with a lightly built skeleton, a long, strong tail for balance, and carnivorous habits. But their limbs were vertical supports for the body, and they walked quite erect. In this they were mechanically even better than the thecodonts.

Saurischians ("Lizard-Hippies")   The saurischians are divided into theropods and sauropods. The *theropods* followed the thecodont way of life. An early theropod from the Late Triassic of North American was 8 feet long and clearly adapted for fast running over open ground. Some theropods were rather small, like *Deinonychus,* "terrible claw." This small dinosaur was a fast-running bipedal carnivore with savage claws on its front and hind feet. While it is possible that these claws were for grooming a sensual and submissive mate, the reasonable inference is that they were for murderous slashing attacks on contemporary lizards and mammals (Figure 10-1). *Deinonychus* and most other theropods had large skulls, jaws, and teeth in relation to their body size. A few theropods came to be built like ostriches, with a long neck and a small head with weak jaws, but they also had a grasping manipulative "hand" on the front limbs (*Ornitholestes,* for example, Figure 10-2). They may have had an ecology rather like a modern ostrich, eating anything available. Their lack of teeth certainly does not imply that they couldn't have been carnivorous, as any wise old owl can testify.

As well as the comparatively small fast-running theropods, there was another theropod group with large body size and immense power, still bipedal. These were the giant dinosaurs like *Allosaurus* and *Tyrannosaurus,* the largest land carnivores of all time. *Tyrannosaurus* weighed a little more than the largest elephant, but stood 20 feet high and was about 40 feet long. Though heavy, these giant theropods could have been reasonably fast runners and probably relied on impact for killing. They had heavy, powerful heads and short, strong necks, and the teeth were long stabbing devices. Some features of the jaw suggest that the meat was swallowed in large pieces rather than chewed in the mouth. Modern crocodiles do this; they have a churning gizzard containing stones, and food is physically broken down

Figure 10-1   The early theropod dinosaur *Deinonychus* was well adapted for fast running and savage attacks on its prey. (Reconstruction after R. T. Bakker, from frontispiece to J. H. Ostrom, *Bulletin Peabody Museum of Natural History, Yale University* 30, 1969.)

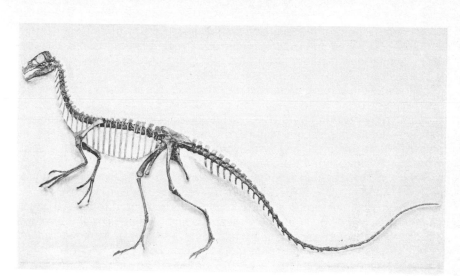

Figure 10-2 Theropods like *Ornitholestes* may have had an ostrichlike biology. (Photograph courtesy of the American Museum of Natural History.)

there. Some dinosaurs have been found with "gastroliths" or "stomach stones" associated with their skeletons, so this was probably a regular feature of their eating habits.

The giant theropods must have depended on large animals for food, and some of these prey species form the second saurischian group, the *sauropods.* While theropods were almost all carnivores, the sauropods were almost all herbivores. They had rather small heads with flattened teeth, indicating a vegetarian diet.

In the Jurassic and Cretaceous, the sauropods evolved into the largest land animals that have ever lived. Their skeletons became massive, and in general they had very powerful limbs, vertebrae, and pelvis, a long tail, and a long neck with a small head. All the sauropods had a permanent four-footed stance, although in most forms the hind limbs were longer and stronger than the forelimbs. This was not simply an inherited defect from the thecodonts; the hind limbs had to support the guts, the pelvis, most of the heavy vertebrae, and the long tail, whereas the forelimbs supported only the rib cage and the fairly lightly built neck and head.

The most famous dinosaurs are sauropods. *Brontosaurus* was about 70 feet long and probably weighed about 30 tons. *Diplodocus* was more lightly built, but reached a length of nearly 90 feet. *Brachiosaurus* was stockier, and its weight has been estimated as close to 50 tons.

Ornithischians ("Bird-Hippies") The second great dinosaur group, the ornithischians, evolved during the Triassic from early thecodonts, although the fossil record of this is not good. There was always a central group of rather small ornithischians, from which other groups evolved at different times. The central group of ornithischians were among the smallest dinosaurs, lightly built and agile like the thecodonts.

Although the ornithischians never reached the size of sauropods, they were the most varied and successful group of vegetarian dinosaurs, occupying the small- and medium-size range during the Jurassic and Cretaceous, and surviving right to the end of the Cretaceous period.

The best-known early ornithischian, *Fabrosaurus* from the Late Triassic, was adapted for an agile bipedal omnivorous way of life, judging from its skeleton and its teeth (Figure 10-3). But in the Early Jurassic the ornithischians began to develop special features of jaws and teeth that were specifically adapted for vegetarian life. These eventually included a large jaw; a large space outside the teeth for a cheek pouch to catch, store, and rechew plant material; "self-sharpening" tooth action; large wear surfaces on the teeth for efficient grinding; and well-packed and numerous batteries of teeth. Because of these adaptations, ornithischians became the dominant terrestrial herbivores of the world for 100 million years, being similar in many of these characters to modern grazing mammals. Only in the largest size range were the sauropods able to outcompete the ornithischians.

The small ornithischians probably browsed on smaller shrubs and bushes, rather like modern deer. Early in the Jurassic, a group of ornithischians developed larger body size and were quadrupedal, slower and more massive in build, often with complicated armor shields. These were the ankylosaurs, which reached 20 feet in length, and were a successful group that survived for the whole Jurassic and Cretaceous. A second group of early ornithischians evolved in a similar way to form the stegosaurs, again up to 20 feet long and armored by plates set to project upwards along the spine.

Another large bipedal group, the iguanodonts, were among the first dinosaurs to be discovered. *Iguanodon* seems to be a simple, large, heavy ornithischian, but this line of evolution led on to the strange "duck-billed" dinosaurs, or hadrosaurs. The "dome-headed" dinosaurs, pachycephalosaurs, are simple, large ornithischians with a cap of porous bone on top of the skull.

Rather late in the Cretaceous the ornithischians gave rise to the horned dinosaurs, the ceratopsians. They are very interesting because of their resemblance to rhinoceroses, although most were as large or larger than the largest rhinoceros. They had a horny beak, one or several horns projecting forward from the skull, and a large "frill" or "collar" extending backward over the neck region. These were certainly adaptations for cropping vegeta-

Figure 10-3 *Fabrosaurus* was an early ornithischian dinosaur, probably agile, bipedal, and omnivorous. (After R. A. Thulborn, *Nature* 234, 75, 1971, copyright by Macmillan Journals Ltd.)

tion and for defense, and the frill also acted as a base for powerful chewing muscles operating the jaw. *Triceratops* was one of the last dinosaurs, and *Protoceratops* is famous because completely preserved nests of its eggs have been found in Cretaceous rocks in the Gobi Desert in Mongolia (Figure 10-4).

RECONSTRUCTING DINOSAUR BIOLOGY

These, then, were the dinosaurs. What kind of lives did they lead? In terms of food, the vegetarians of all sizes must have consumed huge quantities of plant material. Their teeth, particularly in the ornithischians, were well adapted for browsing and chewing, and in many cases they also had grinding stones in a kind of gizzard. Modern elephants can lay waste trees and plantations; the impact of a group of 30-ton dinosaurs on Mesozoic forests and trees must have been immense.

These vegetarians would have been kept in some kind of ecological balance by dinosaur predators, theropods of all sizes. The theropods show signs of being fast, fairly agile animals with powerful head, neck, and jaws. But even at a weight of 10 tons or so, the largest theropods would have faced a heavy problem in trying to kill a 30-ton herbivore, especially one equipped with effective defenses. Some herbivorous dinosaurs had structural protection and often powerful weapons. Ankylosaurs, stegosaurs, and ceratopsians had armor on the head and/or along the spine, and stegosaurs and ankylosaurs had spines or clubs as part of a powerful tail. Probably all herbivorous dinosaurs used the tail as a powerful weapon, just as a modern crocodile does.

Among large modern animals, full-grown elephants, hippos, and rhinos

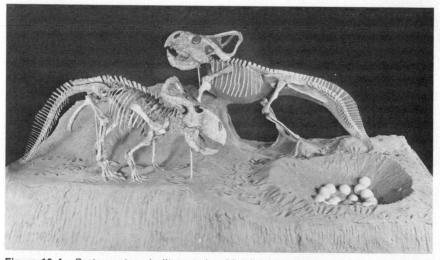

Figure 10-4 *Protoceratops* built nests in which it laid its eggs. This reconstruction is based on material collected in the Gobi Desert, Mongolia. (Photograph courtesy of the American Museum of Natural History.)

are practically safe from carnivores as long as they are in good health. Carnivores like the big cats and packs of dogs and hyenas tend to concentrate on smaller prey species, or young individuals of the large species. There is no reason to doubt that the great carnivorous dinosaurs preferentially preyed on the young herbivores (say those up to 3 or 4 tons!) which they could easily overpower, instead of facing a powerful and dangerous 30-ton adult.

The tremendous size and weight of dinosaurs has led to the argument that some of them, particularly the sauropods, could not have supported their own weight on land; and that they must have had an aquatic way of life, living in swamps. This is probably quite wrong. It is partly based on the idea that sauropod limbs were sprawling, with leg bones projecting sideways from the body. Certainly a sauropod with this kind of limb could not have supported its own weight for very long, because it would be doing a continuous 30-ton push-up. Yet this is the attitude in which many dinosaurs have been reconstructed in museums, dioramas, and textbooks (Figure 10-5).

Dinosaurs should in fact be reconstructed with erect limbs, like the modern elephants and rhinos which they resemble (Figure 10-6). There is then no problem in supposing that sauropods could be supported on dry land by their own erect limbs; and in fact they would have to live in dry conditions or they would be mired in mud. An examination of sauropods by Robert Bakker shows that they had elephantlike feet, limb proportions and joints, and rib cages, all of which are quite unlike those of the swamp-dwelling hippopotamus. There is thus every indication that the ecology of sauropods included a savanna-like open-forest habitat like that of modern elephants.

Figure 10-5 *Triceratops* was a later ornithischian dinosaur, with an appearance and perhaps a biology like a modern rhinoceros. (Photograph courtesy of the American Museum of Natural History.)

A

B

Figure 10-6 (*a*) This triceratopsian dinosaur is miserable because its limbs have been reconstructed as if it were a giant lizard; it is slow and sluggish and movement is clumsy and difficult; (*b*) this triceratopsian is happy because its limbs have been reconstructed by comparison with large living herbivores. Movement is easy and graceful. (Drawings from R. T. Bakker in *Discovery* (Peabody Museum of Natural History, Yale University) 3(2), 1968.)

Other arguments about the swamp habitat of sauropods are not valid; we have seen that their teeth were supplemented by gizzard stones, and there is a good chance that their theropod predators could swim as well as they could, so that a swamp would be no refuge. The long neck of sauropods has been compared to a snorkel tube for breathing while submerged. This would have been quite impossible, for the rib cage could not have operated against the pressure of 30 feet of water. (Ask any scuba diver.) Robert Bakker points out that when sauropods are reconstructed in their correct posture, with erect limbs, and in their correct habitat, the open forest, their long necks would have been an ideal adaptation for grazing on high branches (Figure

10-7). Elephants use their trunk and tusks, and giraffes have long necks, for the same purpose.

The only remaining possible aquatic dinosaurs are the "duck-billed" forms. Even here, John Ostrom of Yale University believes that we have been mesmerized into believing in the duckbill as an adaptation for water life, and he also believes that these dinosaurs were also inhabitants of land environments, rather like gigantic ostriches.

Robert Bakker has gone further in his interpretation of dinosaurs. Erect limbs, as we have seen, are associated in modern vertebrates with high body temperatures and activity levels. Bakker suggests that dinosaurs were warm-blooded and active animals, despite their reptilian nature and their bulk. In fact this makes sense, because the very size of dinosaurs would even out temperature variation because they would take so long to heat up and cool down. If their metabolic level was fairly high, they would have been warm-blooded even if they did not have the very sophisticated temperature control system that modern mammals have. In fact, it is even possible that dinosaurs had more trouble cooling off than they did heating up.

Mechanically speaking, Bakker's idea makes sense too. There is a mural in the Senckenberg Museum in Frankfurt, Germany, showing how it is mechanically possible for a huge iguanodon to break into a trot, and finally a full bipedal sprint, by leaning forward and using the tail as a balance. That this mechanical ability is present shows that it was selected for, and that it was not only used but was important for the iguanodon, presumably in emergencies (Figure 10-8).

There is even a suggestion that some dinosaurs had a very advanced social ecology. The many different horn types and frill types may suggest that they were used for social signaling among a species, just as a bewildering variety of horns and antlers has evolved among living antelope and deer for that reason. Some lizards have multicolored frills round their necks for display in the breeding season.

Figure 10-7 Reconstruction of the sauropod dinosaur with the longest neck, *Barosaurus*. (After R. T. Bakker, *Discovery* (Peabody Museum of Natural History, Yale University) 3(2), 1968.)

Figure 10-8 Diagram to show how a massive dinosaur would still have been well balanced to break into a trot and then into a fast run. (Inspired by a mural in the Senckenberg Museum, Frankfurt, West Germany.)

"Dome-headed" dinosaurs had a skullcap of bone with air spaces in it. Peter Galton has suggested that they evolved It for the same reason that living sheep and goats did—to protect the brain during fierce fights between males in the breeding season. Skulls like this are useful only in cases where the head is pounded against a hard object like another head. It is very difficult to think of any alternative to Galton's idea, so that we have seriously to imagine head-pounding contests between male dinosaurs, each weighing many tons!

The idea of high body temperature and high activity level helps to explain other features of dinosaur biology, and so meets one of the best tests of a new idea. Dinosaurs often have large air spaces within the skeleton, especially the spinal column. (The air space in the spine of *Stegosaurus* was often explained as a kind of second brain cavity!) These air spaces are also found in bird bones, and while they do help to lighten the skeleton, they also aid in cooling during periods of activity.

Dinosaurs did not have any insulating hair, as we know from rare finds of preserved skin. Very small warm-blooded animals need insulation, because they would otherwise lose heat very fast and die of cold. Thus birds and small mammals both havc good insulating coats. But small (young) dinosaurs would be faced with a heat-loss problem if they were really warm-blooded. Since we find that there are no dinosaurs that weighed less than 20 pounds as adults, and since we find that dinosaur eggs are large in comparison to body size, the hatchlings would already be quite large, with fewer heat-loss problems. Large dinosaurs would need no hair—rhinos and elephants have practically none when adult.

In contrast with dinosaurs, the mammallike reptiles and the early mammals had sprawling limbs. In living monotremes—the duck-billed platypus and the spiny anteater—the limbs sprawl. The body temperature is poorly regulated, especially in heat loss, and averages only about 28 to 30° C. Primitive living *placental* mammals thermoregulate, but only near 30° C. Mammals do not seem to have developed erect limbs until the Cretaceous period. All known Jurassic mammals have sprawling limbs. Altogether this gives a clear picture of Mesozoic mammals as small, sprawling animals, thermoregulating poorly, with low activity levels.

Most living primitive mammals are nocturnal, and the mammals as a whole show descent from nocturnal creatures. Hearing, smell, and touch (by whiskers) are night senses that are better developed in mammals than in reptiles, and mammals do have hair insulation against heat loss.

In short, this is circumstantial evidence that Mesozoic mammals were night animals. They would have found it difficult to compete with dinosaurs for most ecological opportunities on land. In the Mesozoic it seems that it was the mammals that were cool-blooded, sprawling, and subordinate, while the dinosaurs were active, warm-blooded, and dominant. Since that time the relative habits of the reptiles and mammals have completely reversed.

Paradoxically, it seems now that the correct idea of the biology of dinosaurs is the one enthusiastically presented for so many years by Sir Arthur Conan Doyle, by horror movies, and by comic books: the professional paleontologists of the last century have been largely mistaken. It is only fair to say that some scientists do not agree with this new interpretation of dinosaur biology, and as this chapter is written a storm of controversy is blowing up. Over the next few years this question ought to be settled as new facts about living and fossil animals are fed into the arguments.

The remaining problem of dinosaur biology is to account for their extinction at the end of the Cretaceous. This will be discussed after we look at world biology during the Mesozoic (see Chapter 13).

SUMMARY

Mammallike reptiles were the dominant land animals in Permian times. But in the Triassic the other reptiles, particularly the thecodonts, outcompeted them. A change in climate may have been partly responsible for this, but the thecodonts also evolved semierect limbs which made them more mobile and more active than the mammallike reptiles. Dinosaurs were descended from thecodonts, and had fully erect limbs. They were probably active and possibly warm-blooded animals, totally dominating land environments and reaching gigantic size. At this time mammals were tiny and had sprawling or semierect limbs; they must have been comparatively inactive and possibly nocturnal creatures.

ELEVEN

THE EVOLUTION OF FLIGHT

FIRST FLIERS

The first land organisms, and the first aerial organisms, were plants. As plant reproduction adjusted to the problems of life in air, spores were evolved that had dry, waterproof covers instead of damp slime as in seaweed spores. Dry spores could then be spread by wind. By Devonian times, plant spores were numerous and widespread enough to be used as zone fossils for the study of Devonian rocks.

Insects evolved flight in Carboniferous times, and a great variety of wing types is found among late Paleozoic insects. Mayflies and dragonflies are living survivors with a very early type of wing that cannot be folded up between flights. This problem had been solved by late Carboniferous times, and most later insects could flex the wing bases so that the wings folded back over the body. It is not known how insect wings first evolved, but insects had a monopoly of the air for 100 million years or so.

Flight gives many advantages, particularly in mobility, though it demands high energy output. Many insects could gather ample supplies of food from the plentiful plants of the Carboniferous coal forests. Half of all known Paleozoic insects had piercing and sucking mouthparts for eating plant juices. In turn, these insects were a food source for the savage dragonflies.

Vertebrates did not take to the air until Triassic times when gliding lizards evolved. Several living vertebrates can glide to some extent by stretching out skin surfaces between their limbs, or in other ways. These living gliders include the "flying" squirrel, "flying" phalanger (a marsupial), "flying" gecko, "flying" frog, *Draco* the "flying" lizard, and even a "flying" snake. All these living animals are found in thick tropical forests. Presumably such adaptations arise in animals that habitually climb trees and jump from branch to branch or from branches to the ground. Any method of breaking the impact of the jump, or of leaping longer distances, would be advantageous and could have evolved very gradually.

The very earliest lizards included some very specialized types. For example, *Icarosaurus* from New Jersey was a Late Triassic lizard which was normal except that its ribs projected very far sideways instead of curving downward in the chest wall. *Icarosaurus* has been very carefully analyzed by E. C. Colbert, and he has shown that when the ribs were fully extended sideways, the skin covering them would have made a perfect aerofoil. Muscles and ligaments between the ribs could have given expert and precise control of the gliding surface. At rest, the lizard could have swung its ribs backward on joints, and folded them up against the body (Figure 11-1).

The Triassic gliding lizards were only about the size of squirrels. We have not yet found their ancestors, and they left no descendants. In many ways they were as versatile as the larger gliding lizards that evolved later; *Icarosaurus* probably had better control over its gliding surfaces, and it still had four perfectly good limbs for walking, climbing, grasping, and manipulating while on the ground.

FURRY PTERODACTYLS

The pterosaurs (or pterodactyls), the most famous gliding reptiles, evolved in Early Jurassic times from some thecodont ancestor. They had a very large skin membrane stretched between the arm, body, and leg, very much like a

Figure 11-1   The skeleton of *Icarosaurus* shows how its ribs were extended into an aerofoil. (After E. C. Colbert.)

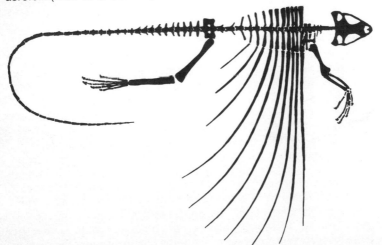

living bat (Figure 11-2). But where bats support the wing membrane with the whole hand, the pterosaurs used only one finger, which left three fingers comparatively normal in structure.

Pterosaurs had wingspans ranging from a few inches to more than 50 feet; giants like *Pteranodon* are the largest flying animals ever to evolve. All pterosaurs had a very light construction, with hollow bones. Even *Pteranodon* probably weighed no more than 40 pounds. Pterosaurs had long heads and jaws which sometimes had many teeth. They are divided into two groups: those with a long tail and those without. The tail probably served as a rudder in the smaller early pterosaurs, but perhaps as flight control improved the tail became unnecessary extra weight.

Much paper has been used for discussing the biology of pterosaurs. In 1974 Cherrie Bramwell and G. R. Whitfield of the University of Reading, England, published the first *engineering* analysis of the skeleton of a pterosaur, the giant *Pteranodon,* and we now have a more exact idea of how they operated. As in many other pterosaurs, the great area of wing membrane and the very light construction of the body suggest that *Pteranodon* was a glider by preference, and would only have been able to fly by flapping its wings in occasional emergencies.

In the air, *Pteranodon* was a more efficient glider than any bird or man-made machine, although its flying speed would have been very low (best performance at 17 mph, maximum speed about 30 mph). In other words, *Pteranodon* was designed for gliding in light winds. Baby pteranodons were probably slightly better gliders than their parents, but at a wingspan of 3 feet or so they could only have flown at 20 mph in a steep dive, and would have performed best at about 10 mph.

Pteranodon had a long crest on the back of its skull to balance the long jaws on the neck (Figure 11-2). The crest was very important. First, it formed a lighter counterweight for the long jaws than the heavy muscles that would otherwise have to hold the head up. Second, by turning the head sideways in

Figure 11-2 *Pteranodon* was adapted for gliding over the sea in search of fish. (After Bramwell and Whitfield.)

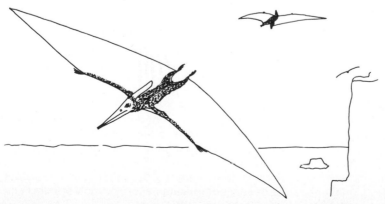

flight, the crest would have acted as an efficient airbrake, either for maneuvering or for coming in to land!

Pterosaurs must do more than fly efficiently; they must eat and they must live on land. It looks as if most were fish-eating animals, probably gliding just above the waves and spearing fish on the long jaws. One Jurassic pterosaur had a 6-inch jaw with needlelike teeth projecting outward. It is possible that some landed on the water and fished from a floating position, though they would then be vulnerable to attack by large marine animals. There would be very little problem in take-off from the sea, because pterosaurs can fly at 15 mph; in a normal sea breeze, they could have taken off simply by spreading their wings. In addition, the Russian paleontologist A. G. Sharov has discovered pterosaurs with *fur* preserved on their bodies! Given any reasonable amount of body oil, the pterosaurs would have been water-shedding, and would have been able to take off from a water surface practically dry.

The hind limbs of pterosaurs are very like those of bats, and must have been practically useless for walking. At best, pterosaurs would have been able to haul themselves along on the ground, sliding on their fur. Like the hind limbs of bats, those of pterosaurs seem best designed for grooming their fur and for hanging upside down from a perch. A 25-foot wingspan has to be folded up on landing, and the pterosaurs would have been able to do this. If the wing was as elastic as a bat's, it would have stowed neatly over the back.

Bramwell and Whitfield envisage the pterosaurs as living like many modern sea birds, in large aggregations on sea cliffs, taking off each suitable day for a fishing expedition over the sea. Landing on the cliffs would not be too difficult because of the low flying speed of the animals. The main hazard would be strong winds—obviously a *Pteranodon* caught in a strong offshore wind would be doomed unless it landed on the water and rode out the storm while waiting for a wind to take it home again. Most pterosaur fossils occur in sediments laid down at sea, and they may represent animals blown too far from land.

If all pterosaurs were furry animals, they would have been warm-blooded. Their brains are quite large, and their biology may have been quite intricate. It is even possible that young pterosaurs, like young bats, were carried around clinging to their parents' fur!

Pterosaurs became extinct at the end of the Cretaceous period. If the climate of the world became more seasonal at that time, with stronger average wind speeds, it is possible that a way of life based on a very large but very light gliding structure simply became impossible.

FEATHERY BIRDS

Living birds are very complex. They are warm-blooded with an efficient heat regulation system that keeps their temperatures higher than our own. They breathe more efficiently than mammals, pumping air through their lungs rather than in and out. They have better vision than any other living animals. Above all, birds have developed the ability to fly better than any other vertebrates, and flight demands very complicated energy supply systems and sensing devices. A bird brain is actually quite large.

Living birds are very varied, ranging from ostriches and penguins that cannot fly to hummingbirds that can hardly walk. But there are enough similarities to show that all birds evolved from one ancestral reptile group. Skull, pelvis, feet, and eggs are so similar in birds and reptiles that Darwin's friend T. H. Huxley called birds "glorified reptiles."

The earliest fossil bird is *Archaeopteryx* from the Late Jurassic of Germany. Several *Archaeopteryx* apparently fell into very soft limy mud, so that they are preserved in exquisite detail, often including the feathers (Figure 11-3). Without feathers, *Archaeopteryx* looks very reptilian; in fact, the first specimen ever found, in 1859, was displayed in a Dutch museum for over 100 years labeled as a pterosaur. The next specimen, in 1861, had feathers and was immediately recognized as a bird, as were three later finds. If it were not for these five specimens from one rock bed, we might think that birds evolved 50 million years later in the Cretaceous, where the next well-preserved fossil birds have been found.

Archaeopteryx had a rather birdlike skull. The braincase was larger than in contemporary reptiles though the cerebellum (coordination center) of the brain was not as large as in living birds. The bones of the skeleton were hollow but did not connect with air spaces as in modern forms. The "breastbone" was not bony, and so the flight muscles could not have been very large. Aerodynamic engineers studying *Archaeopteryx* disagree with one another. It could not have kept up flapping flight for very long, and although it could have glided quite well, its flight would have been rather unstable and it would have had difficulty maneuvering.

Thus, although *Archaeopteryx* is clearly a bird in terms of its feathers, it is really transitional between reptiles and birds in many features. Most likely it evolved from a small theropod dinosaur. We have seen that dinosaurs were probably warm-blooded and must have faced heat-loss problems when they were young or small. The development of feathers would have been the ideal solution to this problem for a small dinosaur, and need not have been connected at all with flight at the time.

There are three main theories of flight in birds, and all of them have to be based on the same data—the fossils of *Archaeopteryx*. They could be called the Running, Jumping, and Standing Still theories.

The first suggests that flight developed in small, fast-running dinosaurs which developed feathers for thermoregulation. The idea is that fast running on open ground could lead to great leaps and bounds, glides, and finally flight as an escape response. The pelvic structure of *Archaeopteryx* makes it clear that it was two-legged and capable of fast running. Actually, this idea is very unlikely. Living fast-running birds on open plains (like ostriches) have reduced wings, because any lift *reduces* traction by the feet on the ground. Aircraft run fast for take-off, but they are not propelled by their wheels. The pilot holds down the aircraft during the take-off run because he gets more power as engine revs increase. Racing cars have airfoils to hold them *down* on the road at high speed. Charles Darwin noticed in South America that rheas (large flightless birds) spread their wings and ran off *downwind* when chased, but this is a different advantage of wings. So there is really no way that fast running on the ground would have led reptiles to evolve flight.

The second theory supposes that small, agile, feathered dinosaurs

Figure 11-3 *Archaeopteryx* is well enough preserved to allow a reasonably complete reconstruction of the body. (Photograph courtesy of the American Museum of Natural History.)

invaded trees, scrambling, leaping and running like living squirrels. Such a way of life would demand agility and coordination, and increased brain size and power of vision. Leaping might naturally have led to gliding and parachuting as in living gliding animals, and finally a crude style of flying might have evolved. At this point in the theoretical sequence we see *Archaeopteryx*.

This theory rests on the assumption that *Archaeopteryx* is actually well designed for scrambling around in trees and leaping and gliding from branch to branch. In fact this is not the case. The hind feet are almost exactly like those of small dinosaurs which are universally accepted as fast-running ground dwellers, and the shape of the claws is not well adapted for grasping and perching on branches. The "hand" is not good for climbing and pulling, because it carries long claws that look more like slashing weapons than climbing hooks. *Archaeopteryx* might indeed have lived in trees, but there are no features of its structure that can definitely be said to show that.

The third theory was put forward by John H. Ostrom of Yale University in 1974. He stresses the close similarity between *Archaeopteryx* and small theropod dinosaurs, which are generally agreed to be active predatory fast-running animals. Ostrom thinks that *Archaeopteryx* too was a fast-running carnivore, catching and eating large insects, lizards, and perhaps even small mammals. The feet have sharp claws, and the jaw and teeth were designed for seizing and piercing prey. The "hands" have slashing claws. Ostrom suggests that the wings developed very powerful downward action to knock down and stun an agile active prey. Hawks, battling against snakes, and swans and geese, can deliver powerful blows with their wings. In *Archaeopteryx*, which is about crow-sized, one can imagine the development of rather precise coordination of wingbeats with the seizure of prey like insects. The wingbeats would also lift the body, and Ostrom suggests that flight could have arisen if the bird developed the habit of leaping upwards after prey from a *standing* position. The final step would be to develop flight as a means of transport rather than a feeding activity. In Ostrom's view, *Archaeopteryx* had reached the evolutionary stage where it was basically still ground-living, but was capable of powerful flapping leaps after prey.

Ostrom's new idea does not rule out the idea that life in trees was an important part of the evolution of birds, but he feels that this happened, if it happened at all, *after* the evolution of *Archaeopteryx*.

LATER BIRD EVOLUTION

After *Archaeopteryx* there is a 50-million-year gap in the fossil record of birds until the Late Cretaceous, when some water birds were preserved. One looked rather like a living diving bird, with powerful swimming feet but small, weak wings; another was rather like a gull, with powerful breastbone and wings.

A. S. Romer has pointed out that a very interesting opportunity arose for birds at the end of the Cretaceous, when the dinosaurs died out. As agile, intelligent, active warm-blooded creatures, birds would then have been very strong competitors with mammals for the control of land environments. In the Eocene of Europe and North America, the giant flightless bird *Diatryma* stood 7 feet tall and had a huge sharp beak, and powerful legs and claws. It must have been a powerful predator among contemporary mammal populations. In South America, *Phororhacos* was another huge flightless carnivorous bird, 5 feet or more tall; it lived from Oligocene to as recently as Pliocene times.

There have always been rather large flightless birds, especially on the southern continents. Living birds like the ostrich, rhea, emu, and cassowary are familiar in zoos, but there were even more interesting forms which are now extinct. The moas of New Zealand reached 10 or 11 feet in height. *Aepyornis* (the "elephant bird") was a huge flightless bird alive so recently in Madagascar that fragments of its eggshells are still found lying loose on the ground. They are unmistakable because they were 2-gallon eggs (probably the largest single-celled organisms ever evolved!). This bird gave rise to the Moslem folktales about the fearsome roc which carried Sinbad the Sailor on its back.

It is always assumed that these birds have "lost" the power of flight. In view of Ostrom's new ideas about *Archaeopteryx* it is *just* conceivable that there is an evolutionary line of birds which has never been able to fly at all.

Penguins have adapted so that they are magnificent swimmers, beautifully streamlined and camouflaged under water, and well insulated with feathers and fat. They are relatively defenseless on land, and so live only on small oceanic islands and in Antarctica, where there are no land-going predators to raid their enormous rookeries. The Cenozoic record shows that penguins have always been penguins, some up to 6 feet tall.

Overall, the fossil record of birds is very poor. They make very bad potential fossils because they tend to be eaten or torn apart by predators and scavengers, or to fall into unlikely places for fossilization. The scanty record is preserved largely because a few birds fell into the sea and were buried in mud fairly quickly.

BATS
The final evolution of flight among vertebrates was in bats. They have a wing membrane stretched between arm, body, and leg, with the fingers of the hand stretched in a fan toward the wingtip (Figure 11-4). The wing is used for flapping flight, not gliding. The hind limbs are modified for hanging from a perch, and there are peculiar features of the pelvis which allow a streamlined small body suitable for flight to have a birth canal for large live young.

The fossil record does not help at all in finding the ancestor of bats, because the first known fossil bat, from Eocene rocks, is already a very specialized form which is totally a bat in all respects (Figure 11-4).

SUMMARY
For millions of years insects were the only flying animals. Gliding lizards evolved in the Triassic period, but they soon became extinct. The pterosaurs were very successful gliding reptiles in Jurassic and Cretaceous times, ranging up to 50 feet in wingspan, and probably furry and warm-blooded.

Figure 11-4 The first fossil bat known, *Icaronycteris*, is extremely well preserved, and its wings can be reconstructed with confidence. (Photograph courtesy of G. L. Jepsen. From the cover of *Science* 154, December 9, 1966, copyright 1966 by the American Association for the Advancement of Science.)

Birds probably evolved from small dinosaurs that evolved feathers as an insulating coat to regulate their temperature. Flight possibly evolved through flapping leaps after prey. The first bird known, *Archaeopteryx,* was still not capable of sustained flight. The great variety of living birds evolved only after Cretaceous times.

TWELVE

THE ORIGIN OF MAMMALS

CYNODONTS

We saw in Chapter 8 that mammallike reptiles dominated the land at the end of the Permian, yet they soon suffered a decline in competition with the thecodonts. The early mammallike reptiles included both herbivores and carnivores, but it was the latter which showed the evolutionary changes leading toward mammals in the Late Permian and Triassic.

Mammallike reptiles still had different ways of life in the Triassic, even though they declined in numbers. There were still herbivores as well as carnivores. *Cynognathus,* a dog-sized carnivore, is the best-known Triassic form. Others were smaller and may have eaten insects and grubs, using sharp pointed teeth for piercing and cutting. A group called cynodonts, including *Cynognathus,* were small to medium in size, with short, stocky bodies and short tails. They were mammallike in that they had a secondary palate separating chewing from breathing, and they had highly differentiated cusped teeth. But a key innovation among the cynodonts was the resuspension of the jaw to hinge on different bones. The main jaw muscle came to attach to the dentary bone, which finally enlarged to dominate the lower jaw as it does in man and all other mammals. Chewing movements of the jaw could become more complex, encouraging a full breakdown of the food before swallowing.

Unfortunately the fossil record of cynodonts is rather poor in Late Triassic rocks, the time that the first mammals evolved from a late cynodont.

A South African animal called *Diarthrognathus* has caused great excitement among vertebrate paleontologists because its jaw has a double joint, one like a reptile's and the other like a mammal's. This sounds impossible because a lever like the jaw can only have one joint. The answer probably is that the jaw had practically no leverage at the joint, but was mostly supported by the jaw muscles alone. *Diarthrognathus* has some features which mean that it is not the ancestor of mammals; a more likely candidate is a South American animal called *Probainognathus,* which also had a "double" jaw suspension.

MAMMALIAN REPRODUCTION

The major difference between living reptiles and mammals is not in their skeletons, but in other biological characters. Reptiles lay large eggs with a large store of energy. The young hatch out as independent juveniles capable of living without parental care. Mammals have small eggs, the young are dependent on their parents, and parental care is well developed. These major differences are accompanied by others. Most living mammals operate at high body temperature and have hair to insulate them, while reptiles are "cold-blooded" and have no hair. Minor differences in the skeleton are not very dramatic, but they can be studied from fossil bones and teeth. In mammals the lower jaw is entirely made up of the dentary bone, and the surplus jaw bones have been modified into the middle ear. Mammalian teeth are highly differentiated, and are not continuously replaced as reptile teeth are. Mammals have large brains, indicated in fossils by the size of the braincase in the skull. How and why did those major and minor differences evolve?

C. A. Hopson has put together some important ideas. Mammallike reptiles were medium-sized, with stocky bodies. They may have had some body hair and some crude kind of heat regulation. Robert Bakker suggests that they may have operated at the fairly low temperature of 28 to 30° C.

But the first known mammals, as identified by teeth and jaw structure, were tiny, much smaller than most mammallike reptiles. Very small animals lose heat very quickly if they are warm-blooded, and to make up for this heat loss they must have very warm coats and they must find and eat large quantities of food. So it seems that the first mammals may have faced very difficult problems because of their very small size (smaller than mice). Almost certainly they still laid eggs as the living duck-billed platypus does, but the eggs would have been very small too, without much yolk, with a short hatching time, and with rather helpless hatchlings.

If small warm-blooded animals have heat problems, their tiny young face an even greater problem. Bird hatchlings are often cold-blooded, with low metabolic rates; they are helpless and dependent on their parents even for warmth. But because they do not have to expend their own metabolic energy on heating, they can put their food intake directly into rapid growth. Full temperature control comes gradually as they grow.

Given this kind of problem for early mammals, there would actually have been an advantage in going all the way and developing very small eggs containing little food which therefore hatched very quickly into tiny cold-blooded helpless young. There would be an advantage in providing a warm, humid environment for eggs and for young, and in having a ready supply of food for the newborn hatchlings.

If early mammals really faced the same kind of problems as modern birds, they probably evolved similar solutions. There is even some evidence for an incubation patch on the belly of some mammallike reptiles, like the brood patch, which is an area of very fine down on birds, where they keep eggs and young warm. A patch more or less implies some kind of nest or burrow for brooding young. It was Charles Darwin who first suggested that an incubation patch could gradually evolve into mammalian breast feeding, in the following way. If a hormone-triggered gland secreted moisture to keep the incubation patch warm and humid during brooding, then a hatchling might have licked the patch for water while the mother was bringing food to it. This might have helped its energy budget in a small way, reducing the number of food-gathering trips the mother was obliged to make, and resulting in increased protection, warmth, and dependence on the mother's presence. If the secretory gland evolved slowly so that nutrition as well as moisture was provided, the young hatchling would have profited even more. Hopson points out here that the presence of breasts only in the *mother* is fundamental to mammals. Male nonparticipation in care of eggs and young was probably inherited directly from the mammallike reptiles. Male chauvinism is nothing new among mammals. On the other hand, male participation in feeding the young is so general in birds that it was probably inherited from their dinosaurian ancestors—which casts an indirect but rather interesting sidelight on the social responsibility of dinosaurs!

The development of suckling can be timed indirectly. Cynodonts had a secondary palate and could chew while still breathing, but even the tiniest young had teeth and so probably did not suckle. But the earliest mammals in the Late Triassic had only limited tooth replacement, and probably suckled in some fashion, even if it was monotreme in style—such as that of living baby spiny anteaters, which simply lick a secretion from pores in the mother's abdomen.

Living monotremes and marsupials hatch or bear their young very quickly; the young are practically fetuses. This is probably similar to Late Triassic mammals. The placenta was apparently a later development among mammals, and it is interesting because it reversed the original mammalian trend toward ever smaller and more helpless young. The placenta prolongs feeding and development of the young *before* birth, just as the large eggs of the dinosaurs gave larger and more well developed offspring (see p. 86). Even today, small mammals have short pregnancies and helpless young, while larger mammals like cattle, horses, and deer have larger young which can run within hours of birth.

This whole story of the evolution of reproduction in mammals is a guess, but a reasonable one. The late cynodonts and early mammals had smaller and smaller body size. If they were warm-blooded to some extent, the rest would follow almost automatically—the similarity with birds that face similar problems shows this. But why were the cynodonts and early mammals pressed toward small body size? If the reconstruction of dinosaur biology by Bakker, Ostrom, and others is correct, it was competition from the thecodonts at a very critical stage in cynodont evolution—after they developed some way toward thermoregulation, but before they had developed a high metabolic rate and an erect gait for direct competition with the earliest dinosaurs. Squeezed between the large thecodonts and the small lizardlike reptiles of the Triassic, the later cynodonts were apparently able to evolve into a habitat suitable for small warm-blooded animals—the night. In doing this the late cynodonts changed so much that they are now recognized as evolving into the earliest mammals. It is hardly surprising that early mammal teeth indicate a diet of insects or other small animals.

EARLIEST MAMMALS

Like the first reptiles, the first mammals are known from very unusual preservational circumstances. The Late Triassic landscape of southwest Britain consisted of weathered limestone islands with deep cracks or joints. Many Triassic reptiles and mammals have been found preserved in the fossil soil and debris in these cracks. The fossils consist mainly of teeth, but already the earliest mammals can be divided into two groups on this basis.

One major group had molar teeth with the cusps in a line. The molars worked by shearing vertical faces past one another in an up-and-down motion, giving a zigzag cut exactly like that of pinking shears in dressmaking. This is very efficient, but it requires very precise up-and-down movement of the jaws. Later, Jurassic descendants had the cutting edges of the molars arranged obliquely, so that the most effective cutting action had a sideways jaw movement as well as an up-and-down motion.

The other Late Triassic mammals had much more complex molars, so that food was trapped, squeezed, and sliced into small pieces from two directions at once as the molar cusps passed one another. The efficient operation of these teeth depended on very precise engineering of the teeth, and coordinated cutting and chewing actions of the jaw. But these mammals were the ancestors of almost all living forms.

Mesozoic mammals were probably all small and nocturnal. Teeth and jaws show that Jurassic mammals were carnivorous or omnivorous. Only a few had molar teeth capable of grinding up fibrous vegetation. In the Early Cretaceous the flowering plants spread over much of the land for the first time. Herbivorous dinosaurs evolved rapidly, and mammalian herbivores also appeared in some numbers. New groups of insects such as butterflies and moths appeared in the new forests, and mammals underwent a radiation based on the invention of an even better type of molar tooth. In Middle Cretaceous rocks one can recognize different sets of teeth which show that marsupials and placentals had become distinct groups of mammals.

So mammals increased in diversity in the Cretaceous, but not in a spectacular way. Yet after the extinction of the dinosaurs at the end of the Cretaceous, the mammals evolved explosively in the Paleocene to large size and great ecological variety. This important event must be seen and explained within the whole world picture, and we shall develop this in the next chapter.

SUMMARY

Mammals and reptiles differ in the structure of their jaws and teeth, and we can see a gradual evolution of true mammals from mammallike reptiles in the fossil record of Late Triassic times. But there are much more important differences between mammals and reptiles, particularly in their reproduction. It seems that mammallike reptiles were being pushed toward smaller and smaller body size because of competition from thecodonts and early dinosaurs. They faced some heat-loss problems as they became tinier, and the features of mammalian reproduction can be seen as solutions to this problem. Birds solved much the same problem in a different way.

THIRTEEN

THE BREAKUP OF PANGAEA AND ITS EFFECTS ON LIFE

THE BREAKUP OF PANGAEA

In the Early Triassic, all the continents were joined together into the giant supercontinent *Pangaea.* The northern continents (Laurasia) and the southern continents (Gondwanaland) were joined at their western end, with the Tethyan gulf between them (see Figure 6-1).

Pangaea was showing signs of stress in the Triassic. Great rifts and volcanic eruptions occurred along lines which are now near the edge of the Atlantic from Nova Scotia to North Carolina. In fact, Pangaea was beginning to split apart.

In the Jurassic, the North Atlantic Ocean began to form as Africa split away from North America. Gondwanaland broke into three pieces which moved slightly away from one another. In other words, the old supercontinent Pangaea really began to break up in an important way.

But the great breakup came in the Middle Cretaceous. The Atlantic Ocean opened up all along its length, and Gondwanaland disintegrated into most of the present southern land masses—South America, Africa, India, Madagascar, and Australia plus Antarctica (Figure 13-1). This tremendous breakup must have completely altered the patterns of ocean currents and the climate of the world. World climate must have become much more "oceanic," with more stable climates both on land and in the sea. Furthermore, the new midocean ridges that were raised in the new oceans between the splitting continents raised sea level quite a lot. Scientists at Columbia

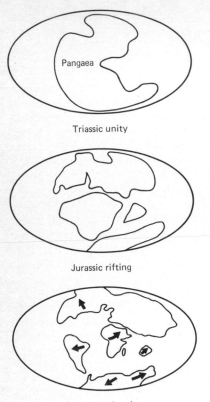

Triassic unity

Jurassic rifting

Cretaceous breakup

Figure 13-1 The break-up of the supercontinent Pangaea.

University calculate that sea level was raised several hundred feet in the Middle Cretaceous in the greatest flooding of the continental edges that is recorded in sedimentary rocks. The events of the Middle Cretaceous must certainly have affected life on land and in the sea. How?

MESOZOIC LIFE IN THE SEA
There are staggering coincidences between the timing of continental movements and the timing of biological events in the fossil record. We have seen how much the geography of the world affected Paleozoic life in the sea, and we must now look at Mesozoic life in the same way. In the Triassic world the great supercontinent Pangaea must have caused a very "continental," seasonal climate. In the sea, suspension-feeding animals like brachiopods, bryozoans, and crinoids had almost been wiped out at the end of the Permian, and Triassic faunas were dominated by deposit feeders, just like Cambrian ones, and for the same reasons. In particular, some animals that had been surface dwellers in the Paleozoic evolved into burrowing ways of life. Bivalve molluscs and echinoids showed a rather spectacular burst of evolution into burrowing forms. As in the Permian, the only refuge for

animals in stable conditions was the equatorial western end of the Tethys sea, and here there was some reef building by sponges and corals. These reefs now form spectacular limestone country in the European Alps.

As Pangaea began to split up in the Jurassic and Early Cretaceous, the climates of the world probably became rather more stable (without the great seasonal variations of a continental climate) and different kinds of suspension feeders evolved. Naturally, the new kinds of suspension feeders were not the same as those that evolved during the older continental breakup in Ordovician times. Mesozoic life is distinctly different from early or late Paleozoic life. But the same ways of life in the sea were followed, ranging from grazing herbivores eating algae, to deposit feeders, suspension feeders, carnivores, and reef builders.

The disintegration of Gondwanaland, the opening of the Atlantic, and the great flooding of parts of the continents in Middle Cretaceous times encouraged the full deployment of Mesozoic life in the sea. Mesozoic suspension feeders included brachiopods, crinoids, and bryozoans, but the major new animals in suspension-feeding habitats were the bivalve molluscs. They radiated into all kinds of sea bottoms and now dominate shelly faunas: deep-burrowing clams in sands and muds, mussels and oysters attached to the bottom by sticky threads or an organic cement, and scallops lying free on the sea floor.

These varied and abundant animals provided food for scavengers and carnivores. Gastropods were herbivores or scavengers in the early Mesozoic, but some of them came to be carnivores during the Cretaceous. A great variety of cephalopods patrolled the sea floor, ranging from ammonites of all shapes and sizes (looking roughly like the living pearly nautilus) to the squidlike belemnites. Most ammonites were apparently stealthy predators and scavengers on the sea floor, while the belemnites may have been wider-ranging like squid. Fishes became very abundant and varied during the Mesozoic, and they more or less completed their development into modern types. Mesozoic reefs were varied; some were formed primarily by corals, some by sponges, and others by huge accumulations of large molluscs.

The most spectacular Mesozoic marine animals, however, were reptiles. Some early reptile groups adapted to water, and their Mesozoic descendants enjoyed tremendous success. Turtles are living survivors; they are actually amphibious, capable of moving about on land, and coming ashore to lay their eggs. In the sea, turtles find abundant food, and probably some degree of safety from predators. Their lives are long once they survive a horrific infant mortality, and their powers of navigation out in the open ocean are legendary. The earliest turtles were Triassic, but later there were huge seagoing forms. *Archelon,* which lived only in the Cretaceous period, was the largest turtle of all time, 11 feet long with a flipper span of 12 feet.

Two major groups of marine reptiles did not survive the Mesozoic. *Plesiosaurs* were reptiles whose limbs had evolved into giant paddles, and the neck and tail had become very long. If they relied on paddling rather than body and fin propulsion, the plesiosaurs probably did not swim very fast. But their bodies were streamlined, and they seem very well designed for slow, efficient sculling through the water, making sudden darts after prey (probably fishes and cephalopods) with their long necks. They had long, wide-gaping jaws with long, sharp teeth (Figure 13-2). An average adult plesiosaur was about 10 feet long, with forty vertebrae in its neck. But some were much larger; the Cretaceous Australian form *Kronosaurus* was 40 feet long, with a skull 10 feet long. *Elasmosaurus,* from the Cretaceous of Kansas, was about the same length, but had no less than seventy-six neck vertebrae.

Some marine reptiles were rather stubbier. *Placodus* was very much like a walrus in size, shape, and ecology; it had hard, flat teeth, presumably for crunching clams dug up from the sea floor. Possibly *Placodus* dragged itself onto land just as a walrus can, to sunbathe, mate, and reproduce.

In many ways the *ichthyosaurs* were the most interesting of all the marine reptiles. Their bodies were beautifully streamlined, so that they had an outline almost exactly like a dolphin or tuna. Fossilized impressions of their skin show that they had powerful fins which had evolved from their front limbs, with a large tail fin and a sharklike dorsal fin. The hind limbs were reduced to very tiny fins. The whole shape was obviously designed for swimming at sustained high speed. Ichthyosaurs had a very long jaw, thin and pointed, with many teeth. Very large eyes sighted right along the jaw line. Obviously ichthyosaurs were very effective predators, much like the living tuna and dolphin that they resemble so closely (Figure 13-3).

But this is not the whole story. Life at sea poses several problems for large animals like these, especially those that evolved from landgoing ancestors as dolphins and ichthyosaurs did. Living dolphins and whales have many adaptations for the problems of breathing, breeding, and caring for the young. Mothers will push their young on the surface until they learn to breathe properly, and they feed their babies milk under high pressure. Ichthyosaurs were reptiles, but they were very highly adapted to a seagoing

Figure 13-2 A marine reptile—a plesiosaur from the Mesozoic. (Photograph courtesy of the American Museum of Natural History.)

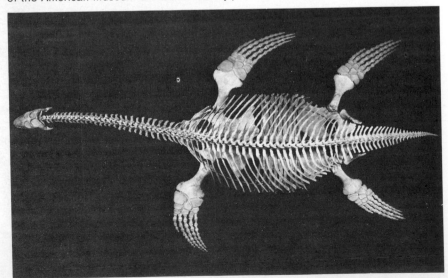

Figure 13-3 An ichthyosaur, a Mesozoic marine reptile known to have given birth to live young. Its similarity to modern dolphins suggests a comparable biology. (Photograph courtesy of the American Museum of Natural History.)

life. Rare fossil specimens show young ichthyosaurs inside the rib cage of adults, showing that they were very unusual reptiles, giving birth to live young.

Both fishes (tuna) and mammals (dolphins) which live like ichthyosaurs are warm-blooded. Their way of life demands constant exertion to keep moving, and this produces body heat. In turn, muscles work best when they are warm and richly supplied with blood. It is no coincidence that tuna has red meat which tastes different from other fishes—it is an animal running at constant warm temperature (chicken of the sea?). Since tuna and dolphins are warm-blooded, ichthyosaurs probably were too, since they were living the same way. Whether this is true or not, there is no doubt that ichthyosaurs were very successful marine reptiles from Triassic to Cretaceous times.

At the end of the Cretaceous, many swimming animals became extinct, including most of the cephalopod molluscs, and the huge marine reptiles. Many other marine animals also became extinct, and extensive changes occurred even among animal groups that survived into the Cenozoic. All these changes at the same time probably could not have been triggered by separate causes, and the problem of their underlying cause deserves a separate discussion (see p. 107).

After the end of the Cretaceous, the place of the cephalopods in the sea was largely taken over by fishes. The giant marine reptiles were replaced by dolphins, whales and seals, and walruses. So the ecology of Cenozoic seas differs from Mesozoic seas only in the kind of inhabitant, not in its major ecological outlines.

MESOZOIC LIFE ON LAND—WHY FLOWERS ARE BEAUTIFUL
Since all organisms eventually depend on plant life for their food, it is important first to see how Mesozoic plants evolved. The early Mesozoic plants provided the basis of the ecosystem into which dinosaurs evolved.

They were descended from the Carboniferous swamp vegetation described in Chapter 8, but as plants invaded drier habitats, they evolved adaptations to retain water and protect their seeds from drying out.

The major advance was the perfection of seeds—fertilized embryos packed in a reasonably watertight coat filled with food for the embryo. The embryo could then survive in suspended animation within the seed until the parent plant dispersed the seeds. Only when the seed was safely transported into favorable conditions would germination occur. The seedling would burst the seed coat and grow, using the nutrition packed into the seed until its new roots and leaves had grown large and strong enough to support and maintain the growing plant.

The plants and trees which evolved this system were related to living conifers. Fertilization in these plants is usually by wind-blown pollen produced in enormous quantities. The seeds may often be packed in cones. Living and fossil relatives also include gingkos and cycads; and Mesozoic forests were populated by tree ferns, and seed plants up to 200 feet high. Naturally fossil beds like the Mesozoic "Petrified Forest" of Arizona contain the large, strong trees, while the softer, smaller, and weaker plants have not been preserved.

In the Early Cretaceous, some plants evolved increased protection for the seed by enclosing it in an extra coat. At the same time, the fertilization process was modified so that many plants came to depend on animals rather than the wind to transport their pollen. This is a very tricky process; somehow the plants must "persuade" the animal to visit them, and then to carry pollen to another plant *of the same species*. The animal must therefore be given some reward for its activity. In some plants an animal is attracted by scent, or by the sight of especially attractive organs (flowers), and its visit is rewarded by a meal of pollen or nectar. In the process, pollen is dusted over the animal—the bee, fly, bat, or whatever it happens to be. The animal, pleased by the experience, is then encouraged to visit another plant with exactly the same scent or flower in order to repeat its pleasure. This will be another plant of the same species, and it will receive some of the pollen carried on the visiting animal and will in turn dust it with its own pollen. To make this system work, each flower should be recognizably different, in appearance or smell or both, from any other plant. In this way a tremendous number of beautiful and varied *angiosperms* or flowering plants evolved, geared toward better protection for their seeds and usually, but not always, evolving more efficient methods of pollination, depending on animals rather than on the wind. (It's usually better to hire a mailman than to float messages in bottles.)

Flowering plants can be recognized in the fossil record in two different ways—by looking for their leaves, seeds, or wood, or by looking for their pollen. These two different approaches have resulted in a tremendous argument about their origin. Paleobotanists who study pollen have discovered that there is no "typical" flowering-plant pollen before Middle Cretaceous times; yet paleobotanists who look at wood and leaves are convinced that "typical" flowering plants occur definitely in the Jurassic and possibly in the Triassic! Daniel Axelrod of the University of California, Davis, believes that both groups are correct in their identifications. He suggests that flowering plants were already evolving from seed plants in the early Mesozoic, probably in stable equatorial uplands where they would be preserved only rarely. At this time, he suggests, flowering plants were pollinated by wind and did not have typical pollen. Later, in the mild climates of the Middle Cretaceous when Gondwanaland fragmented, angiosperms spread widely into the lowlands of the world and became associated with insects that began to pollinate them. Probably this was accidental at first, but developed into a full association with mutual benefit. In Middle Cretaceous times, then, the flowering plants developed their "typical" pollen types, and this is the point at which they appear "suddenly" to the paleobotanists who work with pollen. However, many paleobotanists fly into fits of rage when this problem is discussed, a clear sign that we badly need many more facts to work with.

The flowering plants would have been able to colonize even drier areas than the seed plants, since they had completely enclosed seeds. But more than that, they would drastically have affected the animals around them. For the first time, small animals like insects would have been able to specialize as pollinators, rewarded by pollen and nectar liberally offered to them. This provided alternatives to "stealing" food from plants by piercing their leaves and stems for sap, and possibly running into poisonous chemicals produced for defense by the plants (like nicotine, for example). The Cretaceous development of flowering plants, as well as offering new plants as food for herbivores, set off an explosive evolution of small pollinators, especially insects. These in turn provided extra food variety for insectivores like lizards and small mammals.

Among the dinosaurs, the rise of the flowering plants was accompanied by the evolution of many kinds of new vegetarian types—the horned ceratopsians especially seem to have thrived on the new plants. We shall see that this is important in trying to decide how the dinosaurs became extinct.

EXTINCTIONS AT THE END OF THE CRETACEOUS

Many animal groups died out relatively suddenly at the end of the Cretaceous period. In the sea the large marine reptiles, nearly all cephalopods, and many other marine animals became extinct. On land, dinosaurs died out at about the same time. The flying reptiles became extinct. Was this simply coincidence, or was there some underlying reason? All kinds of explanations have been offered (impacting meteorites, huge volcanic eruptions, bursts of deadly radiation from the sun, selenium poisoning from volcanic ash, Noah's Flood, and so on). But as sensible scientists we should try to find a fairly ordinary set of circumstances to explain our facts before resorting to explanations that would be very difficult to prove or disprove.

A particularly bad hypothesis suggested that the rise of flowering plants somehow triggered the extinction of the dinosaurs (by wearing their teeth away, by giving them constipation or diarrhea, and other such theories), but there is no evidence at all for this. Fossil evidence for attacks of diarrhea would in any case be difficult to find. The fact is that vegetarian dinosaurs flourished on a diet of flowering plants for tens of millions of years before becoming extinct. By offering a new variety of plant species, the flowering plants probably did more to encourage the success of herbivorous dinosaurs than they did to drive them to extinction.

First of all, since marine reptiles as well as dinosaurs became extinct at the end of the Cretaceous, explanations involving events only on land are not convincing. So the rise of the flowering plants, or mammals stealing eggs, or even radiation from outer space, and especially Noah's Flood, would not really affect a plesiosaur swimming deep in a Cretaceous sea. We need a truly worldwide explanation for a worldwide event.

Let us try to see whether the Cretaceous extinctions had any relationship to the movements of the continents. The great spurt of continental breakup in the Middle Cretaceous was not maintained; the separation of the fragments of Pangaea slowed down. The midocean ridges subsided a little as the spreading of the plates dropped in pace. The water which had been pushed over the continental edges in the Middle Cretaceous dropped back into the deeper ocean basins. The effect was not as great as the sea-level drop at the end of the Permian, even though it was caused by a similar mechanism. After all, the continents were separated by now, and the world was not "continental" in geography or climate. But the climatic effect of draining seawater off the continental edges would still have been considerable. Seasonal effects would have become much stronger at the end of the Cretaceous than in the Middle Cretaceous.

It is very significant that the animals that became extinct at the end of the Cretaceous were generally the larger ones. These are the animals that live dangerously in an ecological sense. A dinosaur, for example, was huge and probably lived a long time, eating a vast amount of food. Not many dinosaurs could share the same feeding area, so their breeding populations must have been small. Some fatal interruption like the advent of more severe winters might have played havoc with their juvenile mortality rates, for example, very young dinosaurs were not large and had no insulation against cold. Baby dinosaurs were too large to burrow and hibernate to escape the cold; their parents couldn't either. Obviously one cannot think of all the dinosaurs freezing to death in one bad winter, but a slow change toward a colder winter may have been enough to tip the balance against a slow-breeding and comparatively sparse population.

In the sea the large marine reptiles may have been equally vulnerable to climatic change. The ammonites were beginning to decline toward the end of the Cretaceous, and the climatic change may simply have speeded up their end. It is quite possible that the newly perfected fishes of the Cretaceous seas were outcompeting the ammonites in swimming speed and acceleration, and as a result outcompeting them for food.

CONCLUSIONS

We can now put together the following pieces of evidence:

1. Flowering plants evolved to dominate land floras by the end of the Cretaceous, but this was a gradual process starting perhaps quite early in the Cretaceous.

2. Life in the sea "modernized," beginning in the Middle Cretaceous, and this was followed by a fairly dramatic extinction of cephalopods, large marine reptiles, and many other animals at the end of the period.

3. Dinosaurs became extinct quite suddenly at the end of the Cretaceous.

4. The Atlantic Ocean, and the various parts of Gondwanaland, split open rather quickly in the Middle Cretaceous, so that there was a great flooding of the continental margins and the world became more "oceanic" in character.

5. This was followed at the end of the Cretaceous by a slowing down of continental movements; the seas drained back off the continental edges and the world became a little more seasonal again.

So essentially we can conclude that climatic change, induced in the first place by continental movements, played a major part in causing the biological changes at the end of the Cretaceous. Obviously this story is still rather skeletal, and needs to be fleshed out with solid facts. But it provides the best explanation yet for the Cretaceous extinction which followed so closely after the Middle Cretaceous continental flooding.

Since the Cretaceous the continents have continued to drift apart until now the world is probably about as "oceanic" as it possibly could be. Hundreds of different types of molluscs can be found on tropical beaches, most of them very specialized in their ways of life. There are now more different kinds of organisms in the world than there have ever been, which makes it a very good world for zoologists and botanists to work in. But this could change in the future if the continents were to drift back together again. We are very fortunate to be living at a time when the world is such an encouraging place for specialist organisms.

SUMMARY

The supercontinent Pangaea broke up in Mesozoic times, cracking in the Triassic, separating in the Jurassic, and fragmenting in the Middle Cretaceous. The climate of the world became very "oceanic" and encouraged the evolution of varied and specialized animals in the sea and on the land. In the sea, the newly evolved animals included huge marine reptiles. On land, flowering plants spread widely and encouraged the development of new herbivorous insects, reptiles, and mammals.

At the end of the Cretaceous the continental movements slowed, the world became more seasonal in climate, and there were dramatic extinctions on land and in the sea. The larger animals in particular were very hard hit, and huge marine reptiles, flying reptiles, and dinosaurs all became extinct.

FOURTEEN

CENOZOIC MAMMALS

INTRODUCTION

The end of the Cretaceous period saw so many changes on land and sea that it is also regarded as the end of the Mesozoic era. The Cenozoic, which brings us up to the present, has not been marked by great changes in marine organisms, except that the Mesozoic survivors built up into a very impressive and varied set of animals. The marine fossil record of the Cenozoic is dominated by molluscs, especially bivalves and gastropods—the clams and snails of beach-shell collections.

But on land the Cenozoic was marked by the dominance of the flowering plants and the mammals. The mammals in particular underwent several great Cenozoic evolutionary expansions. Although evolution is the result of all factors operating on organisms, we shall look at two major kinds of response by mammals to Cenozoic events. First, some can be classed as "improvements"—essentially this results in long lines of sequences such as the "evolution of the horse" or the "evolution of the elephant." In almost all these cases, evolution has been acting to fit the animals better for a few particular ways of life, and successive species have disappeared by evolving into a descendant form so different and improved that we recognize it as a new species. Usually these changes can be accounted for in a simple way as long as we understand the biology of the animals. Second, the separation of the Cenozoic continents has resulted in a very strong evolutionary response to changing geography, well worth some study.

EARLY CENOZOIC MAMMALS

All Mesozoic mammals were small and could only play limited ecological roles. But when the dinosaurs disappeared, the Paleocene mammals very quickly evolved to fill their ecological roles, and they also continued to perform the "traditional" Mesozoic mammalian functions as they had for 100 million years.

Most mammalian herbivores are called "ungulates"—these are hoofed mammals that are vegetarian. Although they have evolved similar ways of dealing with plant food, they are not very closely related to one another. Generally, all ungulates have evolved specialized incisor teeth for biting off vegetation, and specialized molars for chewing it. There has to be a large digestive system to handle large quantities of rather low-calorie food, sometimes with special stomach systems for rechewing food, or for fermenting it. Most vegetarians are preyed on by carnivores, so that ungulates are often geared for fast running, with specially designed limbs and hooves. They may also use kicking as a defense, or they may have horns or antlers for defense and for competition within the species.

We naturally find that the first ungulates were small and did not have such specialized teeth as later forms; they also had claws, not hooves, on their feet. But by the late Paleocene the ungulate *Phenacodus* was probably typical of the ancestors of living hoofed mammals; it was a little larger than a sheep, with a rather long tail. The ungulates of this time were very varied, however, and were evolving into all kinds of habits and habitats. In doing so, many eventually became extinct, but others gave rise to the diversity of living ungulates—horses, cattle, deer, rhino, elephants, giraffes, camels, and so on—later in the Cenozoic.

Some Eocene ungulates were huge. Uintatheres were as large as a living rhinoceros, with much of the same body build. *Uintatherium* itself had a large, ugly head with six horns and two large canine tusks. It probably browsed on shrubs and trees, and as an adult it would probably have been safe from attack by any predator. Presumably the horns were for defense of the young, and for competition within the species (Figure 14-1).

Preying on the early ungulates and/or their young were some early mammalian carnivores, usually recognized by their biting and slashing teeth, powerful jaws, and strong clawed feet. There are different styles of hunting, of course, ranging from the stealthy solitary stalk and quick sprint of the cat

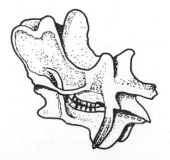

Figure 14-1 The skull of the large early Cenozoic mammal *Uintatherium*. (Drawn from a specimen in the British Museum of Natural History, London.)

to the communal pack-hunting of the wild dog. Yet the carnivores have never had the variety of species that herbivores have had. In fact there have only been two important groups of carnivorous mammals—the earlier *creodonts* and the later true carnivores.

Creodonts evolved as early as Late Cretaceous times, but of course they were small at that time. They had small braincases and their limbs were always rather short and heavy, so that they cannot have been very fast runners. They were the only important carnivores until the end of the Eocene. A typical creodont was rather like a lightly built hyena in size and shape. Presumably like modern hyenas, they could eat either prey that they hunted, or animals that had died or had been killed by other animals. But even a pack of creodonts could not have attacked a large uintathere successfully unless it was already very sick or injured. Like the wild dogs and jackals of the African plains today, their likely prey would have been the smaller early ungulates, young animals, and lizards, frogs, and eggs.

Other likely prey for the creodonts would be the smaller mammals that were essentially continuing their Mesozoic way of life—the insectivores and small vegetarians. Rodents appeared in the early Paleocene, and even the earliest form had the typical large chisel-shaped front teeth. It was squirrel-sized, but either it or a close relative quickly evolved to give a wide variety of animals like the mice, beavers, gophers, and chipmunks that live today. Rabbits look as if they ought to be related to rodents, but in fact primitive rabbits already existed in the Paleocene, and they must have had a long separate history of evolution, based roughly on the same way of life.

Insectivores like shrews were present in the Cretaceous. Living hedge-hogs, shrews, and moles, and the strange and prolific spiny tenrecs of Madagascar, may be rather similar in their habits to their Cretaceous and early Cenozoic ancestors. And it was a shock to find a perfectly preserved bat in Eocene rocks (see Chapter 11). The primates too had already begun to evolve a separate way of life in the latest Cretaceous. Presumably a factor in the evolution of bats, primates, and squirrellike rodents was the relative safety of life in trees, away from the evolving creodonts.

Some early Cenozoic mammals would be easily recognizable to us, for they closely resemble living forms. Others might look very strange at first, but after watching them for a time an ecologist would realize that they were doing things similar to modern animals that he could name—hunting, scavenging, browsing, eating fruit, and so on. In other words, after studying the various early Cenozoic mammals, we see that their ecology was probably much like that of living mammals. But the early mammals have in some cases been replaced by some other forms that performed the activity in a better way. We shall look at some examples of this "replacement-by-improvement."

The Evolution of Horses Horses belong to an ungulate group called perissodactyls (they have an odd number of toes on each foot, at least in all living forms). Other living perissodactyls are tapirs and rhinoceroses. Most of the evolution of horses took place in North America, and it has been intensively studied. Horses have the best-known fossil record of any living mammal. Whole books have been written about horse evolution, although it will be summarized here in a few pages.

The first animal that is recognizable as a relative of horses is a small dog-sized fossil from Paleocene rocks, *Hyracotherium*. It is rather like the early ungulate *Phenacodus* and its relatives, with slight differences in the teeth and feet. The foot was rather rigidly locked so that it had practically no sideways flexibility—in other words, the foot was adapted for fore-and-aft movement only. Thus *Hyracotherium* was already evolving toward a running kind of motion (presumably on firm ground), and its teeth suggest that it browsed on the leaves of bushes.

Hyracotherium lived in North America in Paleocene times, but was much more widespread in Eocene times. It is sometimes called "Eohippus," the dawn horse, but this is not its legally correct name, and it is also slightly misleading because *Hyracotherium* may very well have been the ancestor of tapirs and rhinoceroses as well.

The later history of horses is complicated, as one might expect in detailed study, but several simple facts are plain. Horses evolved toward a larger size during the Eocene and Oligocene. The legs and feet became longer, and all the feet became three-toed. The ribs and back became stiffer and stronger as the animal increased in size and weight. The teeth became more efficient for clipping and grinding up leaves. In other words, early horses were becoming larger and more efficient browsing animals, capable of fast running when necessary. The limbs and feet were not built for agility in dodging a predator, but simply for length and strength to outrun it.

An important event occurred in the Miocene. Modern grasses evolved, and a new food resource became available over much of the world for vegetarian animals that could live on an open plain with nowhere to hide. Horses were well on their way to fulfilling this prerequisite—they had efficient grinding teeth and they were fairly fast runners. Some Miocene horses seem to have forsaken a life eating leafy bushes in the shade for a life on the open grassland—and in doing so they changed a lot. Ecologically, they changed from browsers to grazers.

A few Miocene horses stayed in the bushes and retained the three-toed foot, but a famous horse, *Merychippus*, apparently invaded the plains. It was as large as a modern pony, and although it had three toes, the two side toes were very small—and the large middle toe had a hoof. The teeth were much harder than before, presumably for grinding coarse grasses instead of soft leaves. The later horses show different variations on this theme of successful open-ground living. Just as antelope of various kinds occupy the African plains, different groups of horses, all descended from *Merychippus*, are found in the fossil record of Pliocene times.

One group, whose best-known member is *Hipparion*, remained small and lightly built, and spread over the world everywhere except Australia and South America. At the same time, other horses became larger. *Pliohippus* more or less completed the transformation toward modern horses, because its middle toe was the only functioning part of the foot. It had two main descendants, one of which migrated into South America and gave rise to a large but slow-running horse that eventually became extinct. The other descendant was *Equus,* which includes modern horses, zebras, and donkeys. Strangely enough, horses became entirely extinct in North America after the Ice Age, less than twelve thousand years ago, after they had lived there for 50 million years. Only when the Spanish explorers reintroduced the horse did it return to the continent that saw it evolve from *Hyracotherium* (Figure 14-2).

Horses may seem to have evolved very rapidly, and in fact their rate of change has been quite fast compared to other evolutionary sequences. But Glenn Jepsen of Princeton University has made a very illuminating estimate

Figure 14-2 The evolution of horses. (From G. G. Simpson, *Horses,* copyright 1951 by Oxford University Press.)

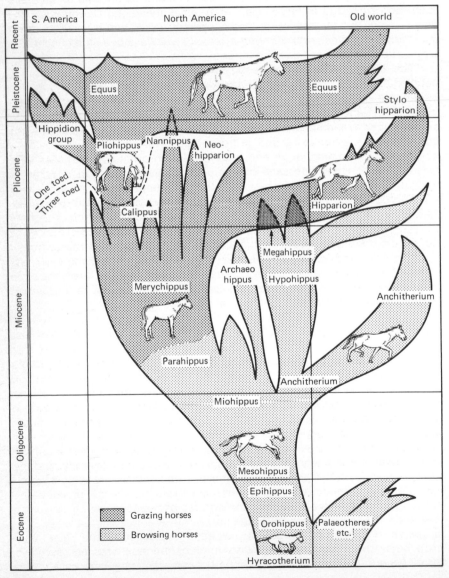

to give an idea how slow this evolution has really been. If all the horses that have ever lived were to march in a solid column, 75 miles wide, at 6 miles per hour, past a grandstand, it would take about 90 years for them all to pass. Obviously it would be impossible to see any discontinuity in the column that would represent the change from one species to the next.

The apparent simplicity of this story of horse evolution is misleading. Many other descendants of *Hyracotherium* which were not horses were highly successful but for one reason or another eventually became extinct— titanotheres are examples. After the large uintatheres became extinct, the titanotheres evolved to large size in the late Eocene and Oligocene. They showed some of the features we have come to expect in large herbivores, after looking at dinosaurs and uintatheres. *Brontotherium* was about 14 feet long and 8 feet high at the shoulder; it was massively built and had elephantlike feet. Males had very large horns fused on the skull between and in front of the eyes; females had horns that were similar but smaller. Again, few predators could have tackled an adult, so that the horns must have been for defense of the young and for competition between males for females. In the long run, titanotheres were unsuccessful, perhaps because their teeth were rather weak and suitable only for soft vegetation.

Rhinoceroses are also descended from *Hyracotherium*. Compared with horses, their history is not well known, but enough has been discovered to show that their evolution has been much more complicated. They generally have been massive, deep-bodied animals with short legs, browsing on vegetation. A few early rhinos were lightly built, with long legs, but they became extinct in the Oligocene, perhaps in competition with early horses. A few rhinos have been very hippolike. But the most spectacular rhino was *Baluchitherium,* the largest land mammal that ever lived. It stood 16 to 18 feet high at the shoulder, and browsed on tall trees. It lived in the late Oligocene and early Miocene in Central Asia; this was for some reason a time when giant mammals were favored by natural selection.

Elephant Teeth Elephants had a complex evolutionary history too. We shall look only at a late part of their history. Modern elephants evolved from late Cenozoic ancestors by changing their chewing action. Pliocene elephants chewed their food in a "normal" way, moving the lower teeth against the upper teeth in all directions. But modern elephants chew only by moving their lower jaw backward and forward in a straight line. Long ridges of enamel on the upper and lower molars meet like scissor blades at a small angle and slice the food to pieces. The new style of chewing is obviously very successful, but it brought with it very important changes in other elephant characteristics. For example, the teeth and jaws changed extensively, and the muscles operating the jaw evolved, which in turn altered the shape of the skull. So an apparently small change in chewing action caused the elephants to evolve considerably in the physical features of the head.

Why Camels Are Ugly Camels do not trot like horses. Instead, they pace, moving both right-hand feet at the same time, and then both left feet. This makes them incredibly difficult to ride, but they can take very long steps, which saves muscular action. Camels are very efficient travelers over long distances on flat ground. Lawrence of Arabia rode a camel on his journeys only partly because (typically) he enjoyed it. Pacing puts a great strain on shoulder, hip, and ankle joints of the animal, because the whole weight of the body sways from side to side with every step. Extremely strong *joints* have evolved in camels (and in their Moroccan riders!) to support this strain, and the neck is set very low on the chest to minimize swaying of the camel's head. So there are good evolutionary reasons for the features which make camels so ugly. In the fossil record we can even see when the change to pacing took place—it was with the great expansion of open grasslands in the Miocene that camels began to evolve the great powerful splayed foot and the reinforced ankle.

EVOLUTION AFFECTED BY GEOGRAPHY

Australia   Everyone has heard of Australia as the continent of marsupials, where the koala and kangaroo have been isolated from the mainstream of mammalian evolution. This is only a part of the story. Australia is different in its plants, insects, and reptiles as well as its mammals. The reasons lie in the changing geography of the region since the Mesozoic. Australia and New Zealand were part of Gondwanaland, and were near the South Pole, joined to Antarctica, when the Middle Cretaceous breakup occurred. Australia and New Zealand did not split from one another and from Antarctica until the early Cenozoic, and only then did they begin to drift north toward their present latitudes. The northward drift of Australia is recorded in the sea-floor sediments round it; for example, it looks as if the westerly currents that flow round Antarctica all year (the "Shrieking Sixties") did not take full effect until the Oligocene (see Chapter 16). This fits with the fact that there is very little fossil evidence on Australia about the history of its animals—the continent may have been very cold until the Oligocene. The earliest Australian fossil marsupials are either late Oligocene or early Miocene, and even then they are rather like living marsupials. So really we have no clues about where Australia got its marsupials. We do know they evolved to fill all the ecological roles that placental mammals fill on other continents; kangaroos are grazers, koalas are browsers, and wombats are large burrowing "rodents." There is a marsupial "mouse," a marsupial "anteater," and even a marsupial "mole." The Tasmanian wolf and the Tasmanian devil are carnivores surviving only on Tasmania; they have been replaced on the main continent by the dingo, a dog which probably reached Australia with man, only very recently.

There once were even more surprising marsupials in Australia. *Diprotodon* was a large four-footed Pleistocene marsupial the size of a rhinoceros. *Thylacoleo* was another large Pleistocene form, called the "marsupial lion." It was the size of a lion, and had very efficient cutting teeth, but it is not clear whether it used the teeth for slicing up large living prey, or whether it peacefully sliced up fruit.

It is usual to talk of marsupials as primitive and inferior to placental mammals, and in many cases it is true that marsupials have been outcompeted by placentals. But it is not one-sided: for example, a kangaroo is rather ungainly when hopping slowly about in a zoo cage, using its tail as an extra limb in what is really a "five-footed" movement. But at high speed a kangaroo is not only very fast (up to 40 miles per hour) but its incredibly long hops are much more efficient than the full running stride of a four-footed runner of the same weight.

The isolated position of Australia has meant that only very mobile or very adventurous animals (bats, birds, man, dogs) have reached the continent from outside. Obviously, the duck-billed platypus, the spiny anteater, and the mass of marsupials reached it early at a time or by a route such that advanced placental mammals were excluded. No one can agree how this happened, and we will have to know more about the early Australian fossil record before we have an answer. The problem is that marsupials must have reached Australia from the Americas where they first evolved. Some people suggest a migration route across Pacific islands, now sunken; others suggest a polar trek across Antarctica from South America. We really need some facts!

South America   For mammalian evolution, South America is in many ways more interesting than Australia, because we know its history in more detail. South America split away from Africa in the Middle Cretaceous to become an island continent with only slight connections with North America. By the early Cenozoic, South America was completely isolated from any further exchange of animals with other continents, unless they were good island-hoppers, for Central America was not yet above sea level. At that time there was quite a variety of mammals in South America, and they proceeded to evolve on their own, undisturbed from outside for 30 or 40 million years.

The South American mammals divided up the ecological roles just as on other continents, but they were rather incredible animals (Figure 14-3). Edgar Rice Burroughs used some of their names for the fantastic animals found on Mars by that great Virginia gentleman John Carter (in *A Princess of Mars* and several other books).

South America must originally have contained a fair proportion of marsupials, probably like modern opossums. Some of these grew larger and more carnivorous. *Borhyaena* was a wolf-sized carnivorous Miocene marsupial, with canine teeth shaped for stabbing and molars evolved into meat-slicing teeth. It was large, lithe, and strong, and was obviously a successful medium-sized carnivore. Later borhyaenids were larger, one of them being as large as a jaguar, but with saber-tooth canines.

The animals on which they preyed were varied. There must have been the "normal" opossumlike marsupials, and smaller forms like rodents. But South America also had placental mammals which apparently were unable to outcompete the supposedly inferior marsupials. These were early ungulates, relatives of *Phenacodus*, and had reached South America during the Paleocene. They evolved to larger size, just as their North American relatives did. There were many ungulates, ranging from rhino-sized down to rabbit-sized, with large gnawing incisor teeth. *Thoatherium* (the thoat of Edgar Rice Burroughs) had an uncanny resemblance to a horse, with a long face, horselike front teeth, grinding molars, a straight back, and slender legs ending in one toe. Some of its relatives looked like camels.

The ancestors of armadillos, sloths, and anteaters arrived in South

Ungulate

Glyptodont

Marsupial carnivore

Sloth

Thoat

Ungulates

Figure 14-3 For a large part of the Cenozoic, mammals evolved in isolation on the South American continent. Some of them are reconstructed in this diagram. (From Kay and Colbert, *Stratigraphy and Life History*, copyright 1965 by John Wiley & Sons, Inc.)

America in the early Cenozoic. Armadillos and their relatives developed heavy body armor for protection. They are successful animals, opportunistic scavengers, and insectivores. The Pleistocene armadillo *Glyptodon* was very large, about 5 feet long. It had a thick armored skullcap as well as body armor, and had a spiked knob at the end of its tail. Naturally its skeleton was very heavy to support this weight.

Sloths now live in trees, moving painfully slowly and eating leaves. But there were huge ground sloths in South America, including one as large as an elephant. Anteaters are descended from the same ancestors, but have specialized to an amazing extent for eating termites, beginning by tearing apart their nests with tremendously powerful clawed forearms.

Most of these animals probably lived on the plains. Monkeys lived in the forests. They were basically like their African relatives, but evolved prehensile (grasping) tails which they could use as a fifth limb. The South American monkeys fill some of the ecological roles occupied by lesser apes like gibbons in the Old World; for some reason, no apes ever reached South America.

This whole ecosystem, in which marsupials and placentals competed equally, suffered a tremendous shock. Only a few million years ago, in late Pliocene times, the Central American land bridge formed, mainly from volcanic action. Animals were now able to move freely between North and South America for the first time in 40 million years. For some reason that is not understood, the North American mammals were able to invade South America quite easily, and they caused the extinction of most of the South American mammals, particularly the ungulates and the large marsupials. A few of the South American animals survived until after the Ice Age, when they succumbed to the latest northern invader, man. Only a handful of South American animals successfully invaded North America, including the opossum and the armadillo, and the giant ground sloths even reached Alaska. But the encounter between North and South American mammals was unfortunately very one-sided, and caused a great loss to the variety of mammals in the world. It is a prime example of the way in which separation of continents increases the diversity of animals in the world, and how geography affects evolution.

SUMMARY OF GENERAL TRENDS IN CENOZOIC MAMMALS

There are many equally interesting stories in the Cenozoic mammal record. All we can do at this point is to summarize some general features. As we have seen, the Paleocene saw the emergence of a great variety of early mammals, and the early ungulates, especially, took over the herbivorous way of life, with the creodonts preying on them. By Eocene times, South American mammals were diverging along their own evolutionary path, while in the rest of the world more familiar and more modern types of mammals appeared. Rodents, bats, and whales evolved, or at least appeared in the fossil record; lemurlike primates and some recognizable horselike mammals appeared.

In the Oligocene the sea retreated as continental drift slowed a little, and climates became more seasonal. Antarctica began to freeze. Monkeys, mice, and mastodons evolved, with some recognizable modern carnivores like cats. Giant herbivores roamed the northern continents—the titanotheres and the gigantic rhinoceros *Baluchitherium.* Australia was isolated with its marsupial animals.

In the Miocene, grasslands spread at the expense of forest. Camels and horses, cattle, deer, and giraffes spread over the prairies and savannas. Elephants became worldwide. Many varieties of ape were evolving in and around the forests of East Africa. By Pliocene times the grasslands extended still further, and large numbers of horses, cattle, and antelope roamed the plains. The end of the Pliocene was marked by the beginning of severe climates in high and middle latitudes as the Ice Age closed in; we will discuss this later.

FIFTEEN

PRIMATE EVOLUTION

EARLIEST PRIMATES

The earliest evidence of primate evolution is in warm, tropical Cretaceous environments in North America. The first primate, *Purgatorius,* is known only by its teeth—one single tooth from the Late Cretaceous and dozens of teeth from the early Paleocene of Montana. The surprising fact about *Purgatorius* is that its teeth are only slightly different from those of early ungulates. Possibly, then, the primates evolved from early ungulates and not from insectivores (as all older textbooks agreed). However, this evolution must have occurred while ungulates still had clawed feet and not hooves, and *they* probably evolved from some small insectivorous or omnivorous Cretaceous mammal. This only underlines the fact that mammals as a group were evolving extremely rapidly in Late Cretaceous and early Paleocene times, and were taking on all kinds of new ways of life.

The earliest primates were generally like the living lemurs and tarsiers, though we should beware of drawing too close a comparison. The best-known Paleocene primate is *Plesiadapis,* which lived in both North America and Europe, showing that the two continents had not split apart very much at this time. *Plesiadapis* was like a large squirrel in size and habits, and it was probably capable of efficient leaping and clinging among branches. The whole structure of early primates fits with the idea that they invaded trees and were mainly tropical fruit-eaters. The primate combination of excellent color vision, coordination and agility, and complex social interaction, can be seen as molded by this way of life. The evolution of curiosity (useful in searching for food) and intelligence (in making the right response to situations) may in turn be associated with this kind of life. This cannot be the whole story, or birds would have developed similar abilities. Birds are not stupid, but they really are intellectual midgets compared with most primates. Physically, the early primates developed "hands" with fingers capable of the fine manipulation of objects, and the eyes came to be set in the front of the skull to give stereoscopic forward vision. Both these characteristics would have been very useful to small animals climbing trees and picking fruit.

There are several well-known Eocene primates, and generally they are uncannily like the living lemurs of Madagascar in structure, and probably in their biology. Leaping and clinging Eocene primates are best known from Eocene rocks in the Western United States, but similar but not identical fossils also occur in Europe.

The Madagascar lemurs were probably isolated during Eocene times. No other more advanced primates have ever reached the island. Even after 40 million years in isolation the lemurs have not changed in their basic body structure, though they do have very complex societies. Lemurs are adept at the kind of movement known as "vertical clinging and leaping." This is very interesting because the front limbs are used for manipulating and grasping and swinging, while the hind limbs are powerful pushing legs. Obviously, from a style of movement like this, all the ways in which monkeys, gibbons, apes, and man move could easily evolve. By emphasizing arms in movement, the armswinging of gibbons could have evolved. The multilimbed agility of monkeys or the four-footed scrambling of heavy apes on the ground can be reached by stressing all the limbs equally. The bipedal state of man could be reached by accentuating the hind limbs for propulsion and the forelimbs for manipulation.

HIGHER PRIMATES

There are three great groups of higher primates: the New World monkeys, the Old World monkeys, and the hominoids (apes and man). The monkeys must have evolved from lemurlike ancestors at least by the Oligocene, when they were present in South America as well as Africa. There is no evidence that any monkeys crossed the opening Atlantic after the Oligocene, and they have evolved separately on each side until today. New World monkeys now have prehensile tails which they use as a fifth limb, and they have other differences in skull characteristics from the Old World monkeys (four more teeth, for example).

No apes have ever been found in the New World, so it is in the Old World that we might expect to find the fossil evidence showing how apes and monkeys evolved from the lemurlike primates. Strangely, early monkeys are known from only a few places in Africa in Oligocene and Miocene times, and they apparently did not reach Asia until the Pliocene. Apes greatly outnumber monkeys in early fossil beds, and they seem to have spread much more widely over the Old World. It is quite likely that monkeys were simply not very abundant or diverse in the early and middle Cenozoic, and their present success has been a rather recent happening—at least, that is what the fossil record suggests.

The only real evidence of primate evolution during the Oligocene in the Old World comes from Egypt, where a tropical rain forest flourished at the time. Deposits laid down in riverbeds contain thousands of tree trunks, some over 100 feet long. Turtles, crocodiles, and fishes inhabited the rivers, early elephants and hippos lived on lush vegetation, and the primates presumably lived in the trees. Two different monkeys have been found in numbers, together with four different kinds of apes, showing that the two groups had already evolved separately for some time. *Aegyptopithecus* is a fascinating early ape which still had some lemurlike features (small brain and long snout), but it is an ideal ancestor for later Miocene apes and eventually for ourselves. It is possible but not certain that the Oligocene monkeys were specializing toward a leaf-eating way of life, while the apes ate fruit and leaves rather than leaves alone.

Miocene apes are best known from deposits in East Africa dating from about 15 to 20 million years ago. Apelike primates were apparently diverging physically and ecologically into the three groups that are living today—the gibbons, the great apes (gorilla, chimpanzee, and orangutan) and the hominids (ancient and modern men). The early apes seem to have been very versatile in their locomotion: they could walk on all fours using the clenched fist of the hand ("knuckle-walking"); they could use arm-swinging from branch to branch; and they were still small and light enough to be powerful springers and leapers with strong hind limbs. In East Africa they inhabited rain forests on the sides of active volcanoes, ranging into drier and more open plains country.

An early gibbon from East Africa, not quite as well adapted for arm-swinging as modern forms, was distinctly different from other Miocene apes. Similar fossils have also been found in Europe.

Dryopithecus was a great ape. Some species were larger than chimpanzees, and some weighed only 30 to 40 pounds. The lightly built skeleton allowed agility, and the ponderous features of living gorillas and orangs had not yet evolved. Elwyn Simons and David Pilbeam of Yale University suggest that one large African species of *Dryopithecus* evolved into the living gorilla. Even in the Miocene it seemed to prefer upland habitats, and the major change since then has been an increase in bulk and a loss of agility and mobility. A smaller lowland species was probably the ancestor of the chimpanzee, as it was light enough for agile clambering in trees, but could also have been fairly mobile in open country at the forest edge.

Dryopithecus is known in India and Europe as well as East Africa. Presumably the living orangutan is descended from an Asian species, to become the lazy fruit-eater in the high trees of the Borneo jungle. Another probable descendant was the extinct ape, *Gigantopithecus*, which lived in the Himalayan foothills and in south China in Pliocene and Pleistocene times. It may have stood 8 feet high and weighed 800 pounds, and its huge grinding molar teeth show that it lived on very rough vegetation.

Also living in forested environments of India and East Africa was another Miocene ape that had many hominid characters, *Ramapithecus*. Its teeth and jaws show that it ate food requiring more grinding action and less cutting and stripping than the diet of *Dryopithecus*—perhaps nuts, seeds, roots, bones, and meat rather than leaves and fruit? Canines and incisors were de-emphasized, and the jaw became less apelike as this trend continued. The foods mentioned are more easily found in open ground than in the forest, and *Ramapithecus* probably spent more time in the open ground at the forest edge than other apes, perhaps becoming more bipedal in the process. Another fascinating sidelight on the biology of *Ramapithecus* comes from examining tooth wear; the molars are unevenly worn from front to back, showing that they took a long time to erupt and that adolescence was a slow process. This is a sign of delayed maturity, and a longer learning time for the growing hominid.

EVOLVING TOWARD MAN

By 14 million years ago hominids were clearly separate from other apes in teeth and diet, and probably in behavior. But there is unfortunately a tremendous gap in the fossil record of hominids after 14 million years ago. It is not until about 3 million years ago that the next good evidence of human evolution appears. Two East African areas are important—the Olduvai gorge and the Omo River basin, both areas which were open woodland at the time, with plentiful water and game. In South Africa, fossils have been found at several places from about the same time span.

The various South African sites contain species of *Australopithecus,* which was much more manlike than *Ramapithecus. Australopithecus africanus* was lightly built, weighing perhaps 40 to 70 pounds. The teeth and jaws were manlike, heavy in proportion to the skull. The large size of teeth and jaws suggests a strongly vegetarian diet, even though meat may also have been eaten. The brain size was about 450 cubic centimeters (man's is about 1200 to 1400 cc), but that is large compared with an ape. The bones of the body are not as well known as those of the skull, but they show that *Australopithecus africanus* walked upright if not quite erect. The wear on the teeth suggests a life-span of about 20 years, with a maximum of about 40 years. Significantly, stone tools are found with the fossils.

Australopithecus robustus, from beds of the same age in South Africa, mainly differed in being bigger, probably 80 to 140 pounds in weight. The skeleton was correspondingly heavier and stronger, and in particular the teeth and jaws were designed for powerful chewing. The skull was strong to carry the muscles needed to work the heavy jaw. The differences between the two species probably reflect the diet. *Australopithecus robustus* probably had a more vegetarian diet (leaves, berries, roots, fruits, and bulbs) while *A. africanus* was more omnivorous, eating meat among other things. In support of this, antelope bones are found more commonly than one would expect with *A. africanus,* suggesting that it enjoyed antelope meat. Crushing and pounding tools found on the sites suggest that nuts, seeds, or bones might have been crushed before eating.

There are also two species of *Australopithecus* in East Africa, a powerful heavy one and a smaller lighter one. *Australopithecus boisei* (the so-called

"Zinjanthropus") was like the South African *A. robustus* except that its chewing ability was even greater, with very large molar teeth and the facial muscles to operate them. The facial muscles were so strong that the brain case was carried quite far back on the skull, so that the face had practically no forehead at all. *A. boisei* was therefore extremely vegetarian, eating very tough food.

The smaller species was *A. habilis*, perhaps about 4 feet tall. Its chief difference from *A. africanus* from South Africa was that it had a much bigger brain (up to 650 cc). There are stone tools and animal remains in these beds at Olduvai, although it is not clear whether one or both species were responsible for them.

The various South African sites seem to date from about 2 to 3 million years ago, which is a little older than Olduvai (1.75 million years). This would make quite a nice story, with *A. robustus* perhaps evolving into *A. boisei* (becoming more vegetarian), and *A. africanus* evolving into *A. habilis*, with its bigger brain.

Unfortunately, this story is too good to be true. Richard Leakey and an international team of scientists have been investigating the area round Lake Rudolf, at the mouth of the Omo River in northwest Kenya. Specimens of *Australopithecus* range from 2.8 to 1 million years old in that area, and there are both large strong skulls and smaller lighter ones. However, Richard Leakey believes that they are the male and female of a heavy species of *Australopithecus*, and not two separate species. In addition, he thinks that *A. habilis* from the Olduvai gorge is so manlike that it should be given the name *Homo* like ourselves. Thus in his view, there was only *one* line of *Australopithecus* in East Africa, and that was the heavy strong form like the South African *A. robustus*.

In 1972, Leakey's team discovered a skull near Lake Rudolf, in beds that are about 2.9 million years old. It breaks the whole problem wide open, because it is very manlike and seems to belong to the genus *Homo*. In other words, if Leakey is correct we now have evidence that man lived alongside *Australopithecus* in East Africa for two million years or so, and, in fact, that man on present evidence is as old as *Australopithecus*.

The new skull from Lake Rudolf is code-named KNM-ER-1470 until it is eventually given a proper name (Figure 15-1). The jaw and forehead are rather manlike (womanlike, actually!), but in particular the brain is about 800 cc, much bigger than that of any *Australopithecus*. Some thighbones found near are also more like man than like *Australopithecus*. Stone tools are also found nearby, and it is tempting to connect them with the presence of an early man.

Of course, these new discoveries pose all kinds of questions. If man was present as an early species of *Homo* nearly 3 million years ago in East Africa, why is there no sign of him in South Africa over the next million years? Did *A. africanus* keep him out? What is the species *habilis?*—was it *Australopithecus* or *Homo?* Who made the tools at Olduvai, and more important, who made the tools in South Africa?

In trying to answer these questions, we must remember that there were no sudden events in the evolution of man from manlike ape. Whether man

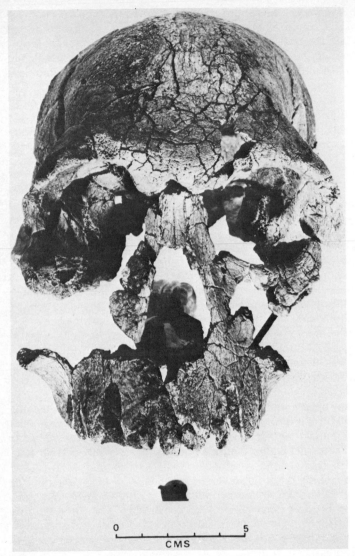

Figure 15-1 Skull KNM-ER-1470 from East Rudolf is the earliest known skull of a man (genus *Homo*), dating from about 2.9 million years ago. (Photograph courtesy of the Kenya National Museums, Nairobi.)

evolved from *Australopithecus* or through some other line from *Ramapithecus,* he did so gradually. At some point in evolution there was no discernible difference between a specimen of *Australopithecus* and a specimen of *Homo,* and at this point the arguments about names become meaningless. At the same time, at this point the fossils are bound to generate furious arguments because they are genuine intermediates and difficult to name.

In fact, what is really significant about the new evidence from Lake

Rudolf is that it suggests man and *Australopithecus* lived side by side in the same area for 2 million years or so—50,000 generations at least—without exterminating one another. Either they were not in direct competition, or they evolved a mutual tolerance which can only be admired in view of man's later conflict with man.

HOMO ERECTUS

Homo erectus was thought for a long time to be the first real species of man. Even if that distinction has been taken away by the new find at Lake Rudolf, he was still the first man to spread from Africa into Eurasia. *Homo erectus* was taller and heavier than *Australopithecus,* and although the skull was thicker and heavier, the teeth and jaws were relatively smaller. The brain was 750 cc or more in the early forms like Java Man, who lived more than 1 million years ago, and had increased to over 1,000 cc in Peking Man (who lived about half a million years ago).

Homo erectus obviously made great progress in the sense that he began to master his physical and biological environment; evidence of bones from Peking Man's living sites shows that he was a successful hunter and had learned to control fire for cooking. Sites of *H. erectus* in Spain and France suggest the presence of deliberately constructed living quarters, and the organization of big-game hunting drives. Both of these show that the social structure of these early men was very complex and stable. Hand axes are the tools that have survived, though undoubtedly there were many other less durable tools made of wood and bone.

There is strong evidence that *H. erectus* was the direct ancestor of *Homo sapiens;* in fact, the only problem is to find the area in which *H. erectus* evolved first into a clearly recognizable *H. sapiens.* Three possible areas have been suggested: Central Europe, Southwest Asia, and East Africa. Possibly *H. erectus* evolved into *H. sapiens* in all three areas: this would be no problem as long as all three advanced populations were interfertile when they eventually met and merged.

Modern man appeared about half a million years ago. His later history, through Neanderthal and Cromagnon Man, belongs in an anthropological text rather than a paleontological one. There is no real dividing line between the two sciences, because the story of evolution is a continuous one; the appearance of *H. sapiens* is only an arbitrary choice at which to end.

The complexity of social interaction at the family, tribe, and district level probably evolved more during this time than the physical attributes of man. R. D. Guthrie of the University of Alaska has suggested that many of the apparently useless features of modern man—facial and body hair, baldness or grey hair in older persons (especially men), skin color, and the behavior patterns of shaving, wearing suits and uniforms, and makeup (and deodorants)—were actually developed by natural selection. He argues that persons who offend other people socially as well as physically are selected against, and that this has built into each different human society a set of physical characteristics and social mannerisms that are as much a part of human biology as walking erect. Social interactions have necessarily become much more complex as human societies have become larger and more concentrated.

Finally, it has been argued that much of human evolution has been affected by the need for a long learning and maturing period (partly, of course, so that the growing juvenile can learn to survive socially in an adult world so full of subtle and puzzling interactions). Along the way, there must have been adaptive value in pushing the learning period further and further into adult life. At the age of twenty two, Alexander the Great had fought four campaigns and won them all, and crossed over into Asia on his way to conquering most of the known world. He did have the benefit of three years studying under Aristotle, but in today's world he would just be graduating from college. David Pilbeam has suggested that throughout man's physical and social evolution, juvenile attributes such as curiosity, learning, imagination, and play have generally been suppressed on reaching the adult stage. Only in a few cultures like ours have they been regarded as generally harmless and perhaps even encouraged. Pilbeam suggests that we are just beginning to use the full potential of the human brain. He had better be right.

SUMMARY

The first primates were probably small tree-dwelling animals, leaping and clinging among branches and eating fruit. This way of life encourages alertness, curiosity, and intelligence, as well as physical characteristics giving agility and manipulative ability. By accentuating arm action in movement, living gibbons have perfected a tree-dwelling way of life. Most monkeys stress all limbs equally in movement both in trees and on the ground. Stressing the hind limbs for movement has led gradually to the bipedal locomotion of man, who uses his front limbs mainly for manipulating objects.

Apes and monkeys had diverged by Oligocene times, and apes and manlike apes had diverged by the end of the Miocene. True man evolved perhaps 3 million years ago, though *Homo sapiens* is perhaps only about half a million years old.

SIXTEEN

THE ICE AGE

HOW ICE AGES BEGIN

To judge from the account of human evolution, one would never imagine that an ice age had gripped the world over the last 3 to 4 million years. In general, the great glacial period which apparently ended about 12,000 years ago mainly affected high and middle latitudes, and did not have a permanent effect on the world's life, although it forced some radical geographical changes on organisms.

Glaciation on a vast scale is not a common event in Earth's history, though it has occurred a few times. There was a very long and widespread ice age at the end of the Precambrian (Chapter 4); a great ice sheet spread over most of North Africa and probably further in Ordovician times. Africa drifted over the pole during Paleozoic times, until there was a small polar ice cap over South Africa in the Carboniferous. Something triggered large-scale glaciation over most of Gondwanaland in the later Carboniferous. Traces of this event (scratched rock surfaces, piles of glacial rock debris) can now be found in South Africa, South America, India, Australia, and Antarctica. There were probably no ice ages after this until the "present" one.

The only external factors that could be important in triggering ice ages are changes in the sun's radiation or in the Earth's orbit. These may indeed have changed, but we have no conclusive evidence at present suggesting that they did cause ice ages. Ice ages are so rare, and at such irregular intervals, that some triggering mechanism here on Earth is probably responsible.

To generate an ice age we need a situation in which Earth's poles are cold and have a lot of snowfall. The sun's radiation would keep the whole Earth at about 45°F if it were all absorbed and evenly distributed. But some radiation is reflected back into space, and the rest is not evenly distributed. The equatorial region receives most, and the poles least—not because of the six-month period of darkness, but because the sun's rays are always at a low angle to the poles. There is a net flow of heat polewards over the Earth's surface as convection redistributes some of the uneven heat by way of warm winds and ocean currents, but these do not warm the poles very much. If anything happens to cut down the free circulation of ocean currents, even this heat flow is interrupted and the poles will become colder. Once cooled down and covered with ice and snow, the poles will reflect back into space most of the little radiation they do receive from the sun. Given enough snowfall, this vicious circle will bring on an ice age as snow and ice thicken and build up into large ice caps.

Earth's present poles show how this might come about. South of Africa, Australia, and South America there is a belt of latitudes (the "Shrieking Sixties") with no land masses. By chance, this is a latitude belt with westerly winds that funnel storms and waves endlessly around the Southern Ocean. No winds or ocean currents penetrate the Shrieking Sixties to bring warm water to the Antarctic, so that it is permanently refrigerated. Slow accumulation of snow has built up a huge ice cap practically from sea level. The South Pole itself stands on 9,000 feet of ice, and the Antarctic holds 90 percent of the world's ice. In this case, interruption of the world's heat flow by *water* has caused an ice cap to form.

In the Arctic the pole is practically surrounded by *land*. The Bering Straits between Alaska and Siberia are narrow and very shallow; the Davis Strait between Canada and Greenland is practically blocked by the Canadian Arctic Islands, and only the narrow sea between Greenland and Norway gives easy access for warm water to the Arctic. The Arctic Ocean is thus practically landlocked, insulated from warm ocean currents, and as a result is covered in the center with permanent pack ice that reflects radiation back into space. The Arctic as a whole is in a very delicate balance, and it would take very little modification of the present situation to generate another northern hemisphere glaciation.

The geographical distribution of land and sea thus controls whether or not Earth's heat is well distributed, and whether or not polar ice caps might form. Plate tectonics controls continental distributions, and the necessary conditions for the generation of an ice age might arise from time to time just by the random motions of the plates forming the Earth's crust.

For example, the deep refrigeration of Antarctica began toward the end of the Oligocene period (i.e., long before the "ice age" we usually talk about). Gondwanaland split apart in the Cretaceous, but Australia and New Zealand did not split from Antarctica until Oligocene times. Not until this final gap opened was there complete oceanic circulation round the Shrieking Sixties, finally insulating Antarctica from warmer ocean water. The Antarctic has thus been frozen for tens of millions of years. Its ice cap froze up more and more water as it slowly built up, perhaps enough to drop world sea level 200 feet.

The climate of the world slowly cooled during Miocene and Pliocene times, and it is inevitable that the freezing of Antarctica contributed to that cooling even if it did not entirely cause it.

The cooling of the Arctic took place much later. The North Atlantic was open in early Cenozoic time, and there is no single event that obviously triggered the great glaciations that began in late Pliocene or early Pleistocene time. Central America was formed between North and South America in late Pliocene times, however. It opened up South America to invasion by North American mammals (Chapter 14), and it may also have altered ocean currents that used to flow between Atlantic and Pacific. If so, it might have intensified the Gulf Stream flowing north into the Atlantic, bringing warmer temperatures, more evaporation, and increasing rainfall and snowfall around the northern end of the Atlantic. Thus it may not be a coincidence that the greatest ice caps in the northern glaciations were centered on Greenland, Labrador, and Scandinavia. This idea is only one of several conflicting theories, however, and it does not explain why there were at least four major advances and retreats of the major ice caps over the last 2 million years— perhaps variations in solar radiation were responsible for these.

There is an important paradox about the world's poles. The Antarctic holds most of the world's ice, yet it is isolated from the great land continents by the same belt of ocean that keeps it refrigerated. It was the fluctuations in glaciation at the much warmer North Pole that produced the advances and retreats of the ice over the last 2 million years.

ICE-AGE PLANTS AND ANIMALS

Obviously the effect of great ice sheets advancing southward over the plains of Europe and North America was extremely severe on the local climate, the plant population, and the animals, and for many hundreds of miles south of the ice sheets. Yet strangely enough the impact was not devastating. Both plants and animals were displaced southwards to warmer zones, but they did not become extinct except in special cases. For example, pollen deposited in lakes shows that at the height of the last glaciation there were tundra plants over most of Western Europe. As the climate warmed, conifer forests slowly spread northward, followed by the temperate mixed hardwood forest that is the natural vegetation in the present climate. But the plants displaced did not become extinct; they are now to be found further north.

Where small glaciers formed in southerly mountain regions (like the Sierra Nevada or the Appalachians), many plants and animals would have been able to escape their effects simply by migrating a short distance *downhill* to warmer places.

Animals tended to show changes as they adapted somewhat to the changing climates. The woolly mammoth and the woolly rhinoceros were much hairier than living relatives, presumably as a response to cold climate. We know about them from cave drawings made by Stone Age man, and also from preserved skins frozen into permafrost in Arctic latitudes in Alaska and Siberia. Huge deposits of jumbled bones are sometimes found in permafrost, and occasionally a complete body with bones, flesh, and skin preserved.

These giant bone-beds have been extensively mined to get at placer gold that sometimes happens to lie beneath them, and they are famous because they have been widely publicized as evidence for huge catastrophes overwhelming not only the individual animals but the whole Earth, too. Instead, they are actually the deposits laid down by ancient spring mudflows as frozen ground gradually thawed after the winter. The mudflows each year flowed downhill, carrying with them to the valley floors any bones and debris lying on the surface. Occasionally an animal would fall through thin ice into the mud and be entombed suddenly and completely.

In Alaska the ice-age tundra was grazed by many large extinct animals, mammoths and mastodons, giant ground sloths, horses, camels, and extinct musk-oxen and bison. Caribou, musk-oxen, sheep, and hares are still living, but some of the ice-age carnivores have also become extinct—the sabertooth cat and a very large bear—although wolverine, lynx, fox, and wolf still survive. The tundra can have been no more hospitable in winter then than it is now, and many of the larger animals must have migrated north and south with the seasons like bison did as recently as the last century, before man interfered with their biology.

PUZZLING EXTINCTIONS

There were many giant ice-age animals that are now extinct. They included herbivores and carnivores, and they lived all over the world. Elephants, ground sloths, and sabertooth cats in North America, the Irish elk (with antlers spanning 11 feet), woolly rhinoceros, and cave bear in Europe, the strange South American mammals, huge marsupials in Australia, moas in New Zealand, and the huge elephant bird of Madagascar (Chapter 11) were all flourishing only a few tens of thousand of years ago. At first it was naturally thought that the climatic changes at the end of the ice age, roughly 12,000 years ago, killed off these animals, and if that was not the direct cause, then ecological disasters set off by vegetation changes must have caused their extinction.

Yet the latest retreat of the ice was the fourth in 2 million years; in other words, all these species had already survived three similar crises. Gradually it became apparent from the fossil record that most of the species had not disappeared at the same time, although those *on any one continent* died out more or less together. The North American species disappeared at about the time the ice retreated, perhaps beginning 11,000 years ago, and so did those in Europe. But the South American extinctions are dated at about 10,000 years ago, the Australian ones about 13,000 years ago, and the African ones much earlier at about 40,000 to 50,000 years ago. Even stranger, the moas of New Zealand were alive less than 1,000 years ago, and so were the elephant birds of Madagascar.

All in all, there is no way that climatic factors alone could have caused these extinctions. There is only one plausible explanation—the arrival of a new ecological factor in the various regions at different times: man, the efficient big-game exterminator.

For the Americas, Paul S. Martin of the University of Arizona has suggested that a band of Stone Age hunters entered North America across the Bering Straits (then dry land) about 11,500 years ago. They followed game south into warmer climates, and found many large animals available for hunting. Martin envisages an immediate population explosion among the invaders, who need not have numbered more than 100 or so at first. The hunters had to keep expanding to the south as they killed off most of the large animals they met. The smaller and more agile animals were perhaps able to escape complete extermination, especially as they would be faster breeders and would have had larger populations to begin with. In fact, it is even possible that the dramatic success of the bison in pre-nineteenth-century times was aided by the extermination of their previous prairie competitors. In other words, the bison eventually got what was coming to them, but later than most.

Martin estimates that the original exterminators might easily have colonized all the Americas within about 1,000 years as they constantly pressed forward after big game. They would be amazed to find the strange South American animals—the giant ground sloth and the giant armadillo—both of which became extinct at this time. Martin's idea involves a southward advance of only about 10 miles a year on average, well within possibility. The extinction dates fit. One need only look at the fate of the North American bison in the late 1800s, or in the killing of about 50 million wild cattle in Argentina within 50 years in the 1700s, to see that Martin's idea is technically within the potential of even Stone Age man.

The extinction of the large marsupials of Australia and the moas of New Zealand coincides in each case with the arrival of Stone Age man, and this is probably the case for Madagascar. The extinctions in Africa seem to coincide with the spread of a new kind of tool, the Acheulean axe. There is enough evidence implicating early man to warrant further investigation of the charge Martin has leveled against him.

In each case it was the largest mammals that became extinct. Probably this is because large mammals tend to have smaller populations than small mammals, and they breed more slowly. They are much more vulnerable to sudden changes in their ecological environment. The large carnivores might not have been killed off by man directly, but might have suffered because man competed with them for game. On every continent, the smaller, numerous, fastbreeding mammals were not usually affected by the extinctions.

Modern man cannot look back at these mass extinctions with any degree of self-satisfaction, for they are still going on, in Asia, Africa, South America, and Alaska. Over the last few years, many of the largest whale species have been driven to a point where they may never recover. Polar bears are shot from airplanes by men who call themselves sportsmen. The emerging African nations are making some progress in setting up wildlife parks; but this policy is in direct conflict with the aims of their land-hungry citizens who would replace the wild ecosystems with imported cattle and goats which are so much less efficient in using the resources of the land. Even the admirable prospect of breeding wild and rare species in zoos is tainted by the methods sometimes used to obtain animals.

It may seem silly to say so, but the world desperately needs the perspective of the paleontologist. We can see the vast expanse of time and the great events that have befallen the inhabitants of the world. Though plants and animals may change by evolution, or by extinction and replacement, their ways of life are ultimately governed by ecological principles that do not change. We have seen whole series of catastrophic changes in organisms, but not in these principles. Evolutionary changes are irreversible, for better or worse, and modern man must think ahead before interfering with the biosphere he does not yet understand.

SUMMARY
An ice age is generated when heat flow over the Earth's surface from Equator to poles is seriously interrupted. Interruptions can come about by the slow movement of plates of the Earth's crust, deflecting or blocking ocean currents or wind belts. The Antarctic was isolated by ocean currents as the southern continents drifted away from it, and it went into deep freeze in Oligocene times. The ice age in northern continents began much later.

The ice ages had surprisingly little effect on organisms, which simply moved to warmer places, or slowly adapted to cold conditions. Many ice-age animals have become extinct, but probably because of man's activities rather than as a result of climatic changes.

FURTHER READINGS

(Figures in parentheses refer to chapters.)

Bakker, R. T., and Galton, P. M. 1974. Dinosaur monophyly and a new class of vertebrates. *Nature,* **248,** 168–172 and references. (10, 12)

Bramwell, C. D., and Whitfield, G. R. 1974. Biomechanics of *Pteranodon. Phil. Trans. Roy. Soc. London, Ser. B,* **267,** 503–592. (11)

Carroll, R. L. 1969. Problem of the origin of reptiles. *Biol. Rev.,* **44,** 393–432. (9)

Clark, R. B. 1964. *Dynamics in Metazoan Evolution.* Oxford University Press, New York. (4)

Colbert, E. C. 1969. *Evolution of the Vertebrates.* Wiley, New York. (7–12, 14)

Gans, C. 1970. Strategy and sequence in the evolution of the external gas exchangers of ectothermal vertebrates. *Forma et Functio,* **3,** 61–104. (8)

Jolly, A. 1972. *The Evolution of Primate Behavior.* Macmillan, New York. (15)

Martin, P. S., and Wright, H. E. 1967. *Pleistocene Extinctions.* Yale University Press, New Haven, Conn. (16)

Orgel, L. E. 1973. *The Origins of Life.* Wiley, New York. (1)

Ostrom, J. H. 1974. *Archaeopteryx* and the origin of flight. *Quart. Rev. Biol.,* **49,** 27–47. (11)

Pilbeam, D. R. 1972. *The Ascent of Man.* Macmillan, New York. (15)

Ponnamperuma, C. 1972. *The Origins of Life.* Thames & Hudson, London. (1)

Schopf, J. W. 1970. Precambrian fossils and evolutionary events prior to the origin of vascular plants. *Biol. Rev.,* **45,** 319–352. (2, 3)

Simons, E. L. 1972. *Primate Evolution.* Macmillan, New York. (15)

Thomson, K. S. 1969. The biology of the lobe-finned fishes. *Biol. Rev.,* **44,** 91–154. (8)

Valentine, J. W. 1973. *Evolutionary Paleoecology of the Marine Biosphere.* Prentice-Hall, Englewood Cliffs, New Jersey. (5, 6, 13)

Wilson, J. T., ed. 1971. *Continents Adrift—Readings from* Scientific American. W. H. Freeman, San Francisco. (5, 13).

INDEX